Honey Bee Drones: Specialists in the Field

ISBN: 978-1-912271-52-8

Published by Northern Bee Books © February 2020

Northern Bee Books, Scout Bottom Farm
Mytholmroyd, Hebden Bridge, HX7 5JS (UK)

www.northernbeebooks.co.uk

Tel: 01422 882751

Front cover illustration by Eunike Nugroho, based on a photograph by Guillaume Pelletier.
www.eunikenugroho.com

Book design by SiPat.co.uk

Honey Bee
Drones

Specialists in the Field   Graham Kingham

Contents

Introduction:
Apis mellifera

The Swedish botanist and zoologist, Carl Linnaeus, introduced the system of nomenclature for identification of life. Linnaeus gave the honey bee the name *Apis mellifera* in 1758 and it appears in his book "Systema Naturae" (10th edition). "*Apis*" is the Latin word for "bee"; "*mellifera*" comes from the Greek "melli" honey, and "ferre", to bear; hence the scientific name means the honey-bearing bee. Three years later in 1761 Linnaeus realised his earlier mistake that honey bees bear nectar not honey, so he attempted to change the name to *Apis mellifica* the honey-making bee. However, according to the International Code of Zoological Nomenclature, the older name has precedence (the Principle of Priority, first formulated in 1842) so we have continued to use the name *Apis mellifera*.

There are three castes of *Apis mellifera*: workers, queens and drones. The worker is about 16 mm long, the drone 19 mm and the queen 25mm. From the time the egg is laid until the final pupal stage the difference in development between the castes is worker 140mg, queen 250mg, drone 346mg, representing 900, 1,700 and 2,300 times the egg weight, respectively. Figure 1 shows a comparison of the caste sizes.

This is reflected in the development time between the worker and drone only.

This little book concentrates only on the drone - the male honey bee. It provides details regarding the drone's internal and external anatomy, production and development, behaviour, role in the hive and in genetics and more. Copiously illustrated, the book also discusses the latest research updates in drones.

Figure 1: Comparison sizes: drone (above) and worker (left)

The honey bee drone

The only known function of a honey bee drone is to mate with a virgin or partially mated queen. They do not collect food, sting, produce wax, or feed the young but do contribute to heating in the hive. Most of the glands found in the workers are absent from the drones. The male honey bee, commonly known as a drone, was supposedly named after the droning noise given off by his wing beats. The Oxford dictionary quotes it as coming from the old English word; drān, dræn meaning 'male bee', from a West Germanic verb meaning 'resound, boom'. He is unique as he has no father but only a grandfather. His major role is in maintenance of the species and thus he possesses no other specialized organs other than reproductive ones. He is a specialist in the field when it comes to mating with a queen, congregating in drone mating areas with other hopeful males. The key role of the drone is to pass his mother's genes on to the next generation; hence the large investment cost into rearing him.

The drone's external anatomy

The key external anatomical differences between the drone, the workers and queens are his size, being about 2.6 times bigger than the worker for reasons which will be explained further on.

Starting at the head, the *antennae* in a drone are larger and have an extra annuli in the *flagellum*. The compound eyes are also much larger, containing three times more lenses, named *ommatidia*, than workers.

The thorax is much larger containing more muscle fibre and is extensively covered in hair. The *abdomen* is squarer at the end and curves inwards to help when mating. The wings reach to the end of the *abdomen* unlike the worker's wings which terminate midway along the *abdomen*. Figure 2 shows the lateral view of the drone.

Figure 2: The drone, lateral view

Wings

The drone's wings cover all of his body down to the tip of the *abdomen* unlike the workers that terminate half way down their abdomen. It is believed that the queen can fold her wings back during the mating process, and the drone's larger wings are used to support them both during the initial stage of mating. One question arises over the reason why drones are so large in comparison to the queen and workers. It has been suggested that the size difference is due to the drone supporting the queen during mating. The force required lifting a drone and a typical virgin queen is 3.4 mN. Drones can produce 3.95 mN under optimal thermal conditions of 40°C. The reverse cannot occur: a queen cannot lift a drone and herself, for the queen at best produces only 2.15 mN of force. Figure 3 shows the dorsal view of the drone.

Some recent filming shows the drone holding onto the queen's abdomen whilst still flying, then repositioning himself for insertion of his *endophallus*, at an angle, forming a V shape when he stops flying and is supported by the queen until the act is over in under five seconds. He then flips over leaving his mating sign behind in the queen's sting chamber.

Thorax

One major anatomical difference between drones and workers is the drones' larger flight muscles found inside the *thorax* together with a thicker cuticle to support them; on the outside the drones also have a lot more hair than workers, much like a mammal's fur. This is believed to aid in keeping the flight muscles warm before taking off. It is estimated that humans have about 100,000 hairs on their heads, whereas honey bees have up to three million on their bodies. Figure 4 shows a lateral view of the drone's *thorax* and head - note the square shape and hairiness.

Figure 3: The drone, dorsal view.

Figure 4: The drone, lateral view of thorax *and head. Note the hairiness.*

Figure 5: Anterior view, the drone's *abdomen.*

Bees generate sound not only through movement of their wings but also with their *thoracic* muscles. Although they use these muscles to move their wings, they can uncouple their wings and use the muscles to produce heat and generate acoustic signals.

Abdomen

The *abdomen* is less pointed than the worker and the drone's external opening has specialised claspers and allows the *endophallus* to pass through.

Figure 5 gives the anterior view looking at the drone's *abdomen* exit area, the *phallotreme*, showing a larger opening through which the *endophallus* everts and the anus exits. By comparison, Figure 6 gives an anterior view looking at the worker's *abdomen*, showing one of the lance tips of the sting protruding out.

The head

The eyes

The compound eyes of a drone are much larger, having up to 11,000 *ommatidium* (worker bees have about 6,300, this is equivalent to 1.75 x) thus enabling them to locate the queen in flight. Furthermore their eyes are positioned in front of the head unlike those of the workers that are positioned on the side, giving a drone near 360 degree vision. Drones have a bigger area in their brain given over to sight. The drone eyes are specialized to select moving objects against a clear blue sky and they show distinct regional specializations. The extremely

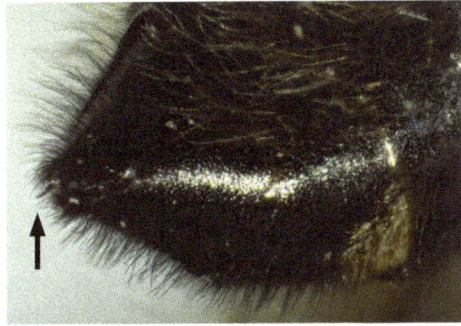

Figure 6: Anterior view looking at the worker's abdomen showing one of the lance tips of sting protruding out.

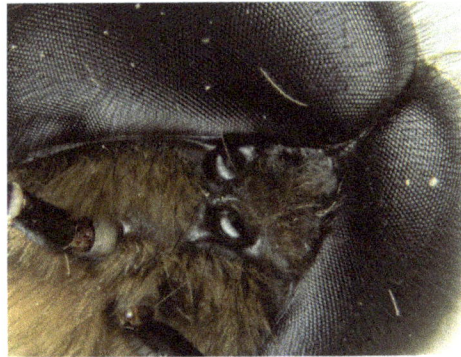

Figure 7: The drone, hexagon-shaped Ommatidia.

Figure 8: The drone, anterior view of the head.

enlarged facets located in the dorsal region of the eye, enable the drone to see a queen at about 50 metres. Figure 7 shows the hexagon-shaped *ommatidium* of the drone viewed from the front.

The light-detecting organs, ocelli, are located on the top of the head and covered with hair in the worker. The drone has his pushed down. This is due to the much larger compound eyes taking up the facial space. They need to be orientated towards the sky for their function of seeing polarised light, which they use to navigate by on cloudy days. Figure 8 shows the anterior view of the drone's head, while Figure 9 gives the anterior view of the worker's head showing the smaller eyes in comparison to the drone.

Figure 9: Anterior view of the worker's head.

Colour Vision in Bees

Bees, like humans, are trichromatic having three receptor types; unlike humans they are sensitive to ultraviolet light, with loss of sensitivity at the red end of the spectrum. This spectral range is achieved by having a cone type that is sensitive to UV wavelengths, and two that are sensitive to "human visible" wavelengths. Because 'colour' is the result of differences in output of receptor types, this means that bees do not simply see additional 'UV colours'; they will perceive even human-visible spectra in different hues to those which humans experience.

Fortunately, as any nature film crew knows, we can gain an insight to the bee colour world by converting the blue, red and green channels of a video camera into UV, blue and green channels. Many fruit, flowers, and seeds contrast with their background much more strongly in UV than human-visible wavelengths.

The mating sign left in the queen is coloured orange to our eyes. However bees see this in ultraviolet light, contrasting and advertising its presence. Bees do not see colour images when flying at speed; just black and white with a flicker fusion rate of 300 per second (compared with 20 in humans as quoted from Dade). They are thus able to see things at speed clearly that to us would be a blur.

The mandibles

The honey bee has two opposing spatula-like *mandibles* attached to its head; they differ between the queen, worker and drone and perform specialized duties in each. Made of *chitin*, they are hardened by being *sclerotized*; this allows them to be extremely strong.

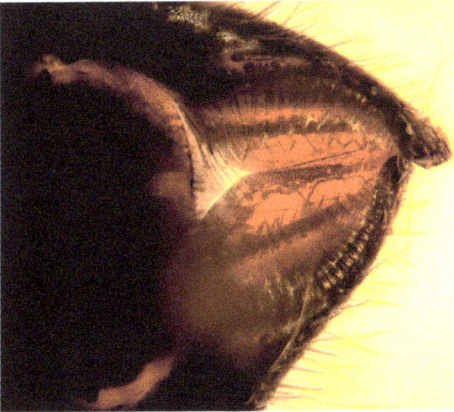

Figure 10: *Worker's* mandibles from beneath showing overlapping jaws (100 x magnification).

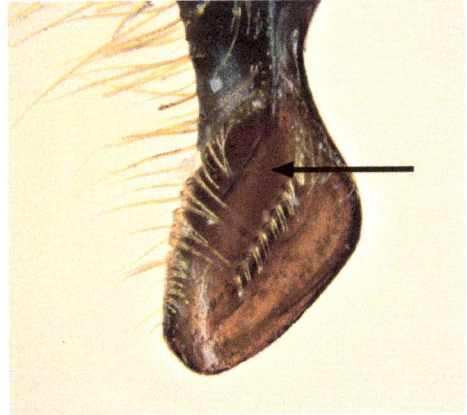

Figure 11: *Workers* mandibles inside view (40 x magnification).

Worker bees

Older worker bees often have saw-like notches at the edges of their *mandibles*, owing to wear and tear. The outside edges have long hairs that overlap; these are innervated and are believed to be mechanoreceptive, sensing biting action. Other smaller hairs that hang down from above over the jaws are also sensitive to movement. Figure 10 is the view from beneath the worker's *mandibles* - showing slight overlap, allowing one jaw to slide inside the other.

The workers are the specialist manipulators. Their first job is to mould wax for the comb with their *mandibles* within the hive or in the bush after swarming. A second role, recently discovered, is to bite the *Varroa* wherever they can! Take a look at the damaged ones on your hive floor, or *Varroa* inspection tray.

The main job of the young workers between 5-9 days is to feed the grubs and queen; they achieve this by producing brood food from the head glands and secrete it down a groove on the internal surface of the *mandibles*. The concave shape and feeding groove is illustrated in Figure 11.

On venturing outside the hive the worker uses her *mandibles* to collect *propolis* which is deposited on her back legs. On returning to the hive the house bees will take off the *propolis* with their *mandibles* and store it onto the edges of the hive or use the *propolis* to seal up a crack. The *mandibles* are also used to support the tongue when drinking or collecting nectar (Figure 12). One other job of the house workers is to remove debris, dead bees and *pupae* from the hive, again with their *mandibles*.

The Queen

The queen has limited but important functions of her *mandibles*; they are toothed with a single cusp at the far end and are used to dig her way out of her queen cell. The outside of the queen's *mandibles* has more hair than those of the workers.

The Drone

The drone has a smaller set of jaws, similar to the queen's; they are also cusped at the bottom and covered in even more hairs (Figure 13). It is believed that they serve no major role for the drone, although like the queen they are used to escape from their cells. Drones and queens lack the feeding groove that workers have as they do not contribute to rearing brood.

Figure 12: Worker's mandibles (40 x magnification).

The tongue

The drone has a tongue about 3mm long as they only have to feed themselves from within the hive stores and not from flowers. On the other hand, the workers have a tongue about 6mm which is required to reach the nectaries of the flowers they visit.

The antennae

The importance of the *antennae*

The word *antenna* (plural *antennae*) comes from the Latin antenna, meaning a horizontal mast spar that was designed to spread a squarerigged sail. Among other uses, the term is used for the feeler organ on the head of an insect, crab or other creature.

The bees' world is governed by *pheromones*, which give them information as to the state of the colony, so smell is a very important means of information and communication inside and outside the hive. *Pheromones* are substances that are secreted to the outside of the

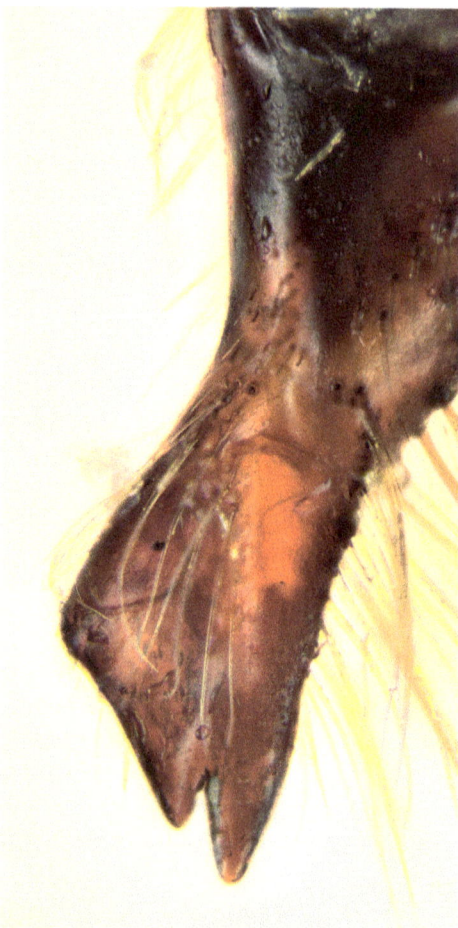

Figure 13: The drone's notched mandibles (40 x magnification).

body, by *exocrine* glands, of which the bee has several. These are picked up by the other bees (and sometimes other insects) and depending on what gland it came from, a specific reaction can result - behavioural, developmental or physiological. Examples of this might be the alarm *pheromones* given off when stinging, queen substance uniting the colony and the *Nasonov* pheromone produced from a gland on the tip of the *abdomen* of workers, and used at the entrance to indicate the hive location to returning foraging family members. All bees are able to smell through their *antennae* and incidentally, feet.

The *antennae* have four segments: the ball and socket joint connecting with the head, the longer *scape* (an upright stem), and then the small, bent top part called the *pedicel* (small stalk), followed by the long *flagellum* (little whip) - the sensory part of the *antenna*, these have 10 *annuli* in the worker and 11 in the drone. Situated on these areas are various hairs and specilised pegs and pits that have porous areas which allow odours to permeate through.

It is within the *pedicel* that the collection of sensory cells known as *Johnston's* organ is situated; it is used for determining air speed when in flight and sound detection.

Drone *antennae* have about double the surface area of a worker bee including an extra *annulus* in the *flagellum* sections on their *antennae* - making a total of eleven (Figure 17 and Drawing D1). Overall they have about 16,000 plate receptors (worker bees have 2,700) which are

Figure 14: The drone, dorsal view of tongue.

Figure 15: The worker, showing longer tongue.

Figure 16: Drone helping itself to honey.

thought to be used in seeking out the queen *pheromone*. It has been found that drones can smell the queen pheromone up to 800 metres away when flying compared with the 50 m for detection by sight. However, the worker bee does score higher on taste

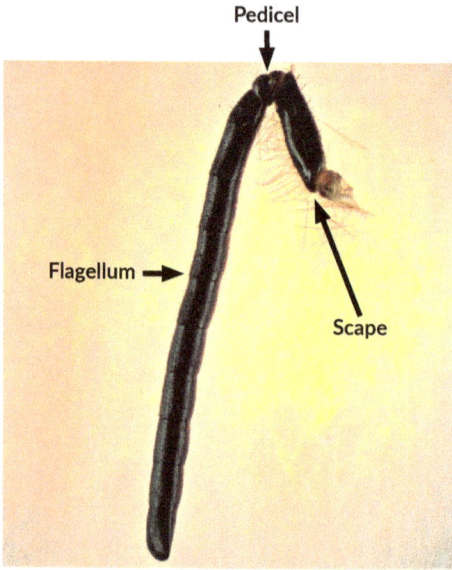

Pedicel

Flagellum

Scape

Figure 17: The drone's antenna (10 x magnification).

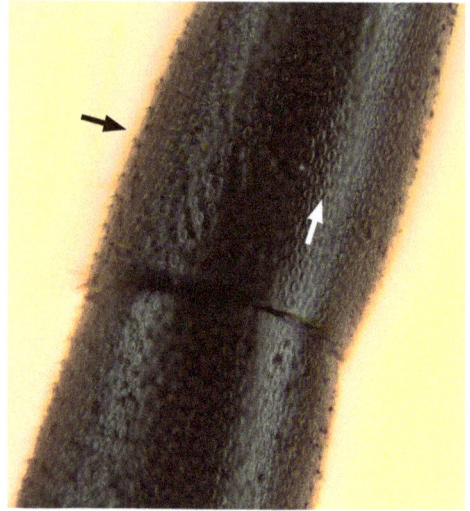

Figure 18: Sensilla on the antennae. White arrow, sensilla. Black arrow, peg like hairs. (40 x magnification).

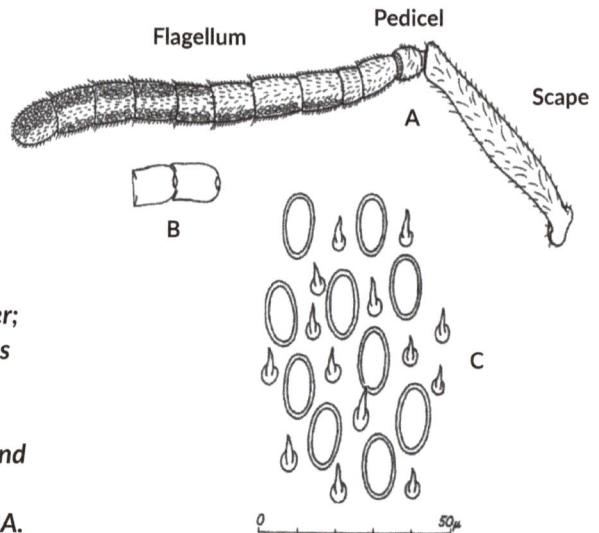

Flagellum

Pedicel

Scape

A

B

C

Drawing D1: A, antenna of a worker; B, jointing segments; C, sense plates and sense hairs on the antenna. Taken from Dade's book: Anatomy and Dissection of the Honeybee. Kind permission to use them has been obtained from copyright holder IBRA.

receptors on the *antennae* - 2,000 against 400 for the drone - reflecting the unique role they both have in supporting the colony.

The *scape* has two muscles internally to move the *antennae* about on its ball and socket joint on the head; the *flagellum* and *pedicel* are kept rigid by blood pressure only. Figure 18 is magnified to show the sensilla.

Figure 19A: The worker bee pollen press and comb (40 x magnification).

There are pitted oval pores and small peg like hairs on the outside. The important ones are mainly found in the eighth *annuli* of the *flagellum*. The *sensilla placoid* are specialised individual oval sense organs connected to nerves which allow the drone and worker to taste. Smell is sensed by the *Basiconic* and *Trichoid sensilla*, the hairy peg like structures; other sensors act as strain gauges, sensing stress and movement such as how far the head will turn. Note the carbon dioxide levels, wind speed and temperature are sensed by other *sensillae* both internal and external.

Figure 19B: The drone's rear hind leg (missing the workers' corbicula).

Figure 20A: Workers foot

Figure 20B: Drones foot

The feet and claws

The three pairs of legs, while similar for locomotive purposes across the three "castes", bear task-related adaptations for the specific functions they will undertake. Thus, a worker is uniquely equipped with a pollen press; they are able to use their front legs to taste with; however the drones appear to lack these sensors and does not collect pollen so has no use of a pollen press. Figure 19A is the pollen press on the hind leg of a worker.

The claws of the worker and the queen are only slightly different in details and outline, although the claws of the queen are much greater in size than those of the worker, but the drone's claws are large and very strikingly different in shape, more angular than those of either the worker or the queen.

The drone's internal anatomy

Function

The drones have a unique mating organ, the *endophallus*, meaning a *'penis* held within'. It consists of a large complicated tube inside of its *abdomen* cavity. In order to ejaculate *semen* the drone must first evert it outside of its body via the opening at the tip of the *abdomen*, the *phallotreme* which also is the exit for the *anus*. There are no muscles or nerves in the *endophallus*. It is everted by muscular pressure that surrounds the abdomen, forcing the *haemolymph* (bee's blood) and air, under pressure, to evert the *endophallus* in two stages.

At the second stage the end of the tube is inserted into the queen's sting chamber, which in mating biology is called the *bursa copulatrix* (mating pouch), and then the second stage of eversion begins, forcing *semen* to be ejaculated at pressure into the queen's ovary ducts.

The *endophallus* is composed of three main parts, the *bulb*, the *cervix* (neck) and the *vestibulum* (cavity). The *bulb* is provided with *chitinous* plates (hardened plate-like structures, to help keep the bulb rigid during eversion). The *cervix* has a *fimbriate* lobe attached mid way (a petal like structure) at one side and transversal folds at the other side. It is possible that the *fimbriate* lobe, inflates with *haemolymph* at its eversion helps to free the mating sign from the *endophallus* and acts as a pressure reserve. See figure 21. Dade gives an analogy of a rubber glove with a finger turned inside out and by blowing through the glove entrance the finger is everted to the outside.

Figure 21: The inflated fimbriated *lobe*.

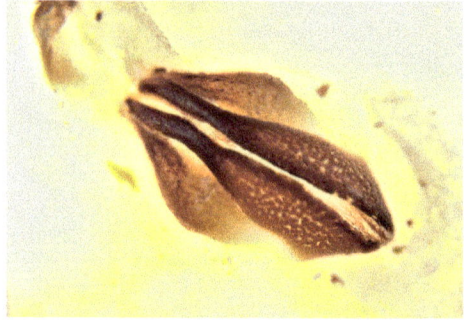

Figure 22: *illustrates the dorsal plate of* bulb, *a structure within* the bulb *that keeps it rigid.*

Other honey bee species have similar organs but they differ in shape at eversion which stops them from mating with other species.

Drawing 2: 1 showing internal reproductive organs. Reproduced by kind permission of Professor Jerzy Woyke 2016.

B–bulb; C – cornua; Ce – cervix; ED – ejaculatory duct; EP – endophallus; MG – mucus glands; T – testes; V – vestibulum; (modified after Woyke 1958a and b)

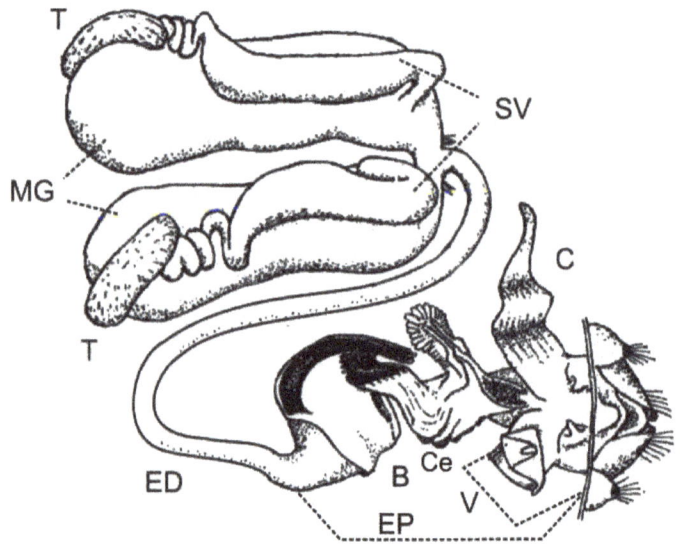

The first completed stage of eversion

The orange colour (shown in Figure 23) denotes that this is a mature drone that is fertile with fully developed *sperm*. It is this part that is visible and left sticking out of the queen's sting chamber after mating, known as a mating sign or signal, and has to be removed by other drones before they can mate. The two orange horns are used to locate onto the queen during mating and leave their residue orange secretion on the previous mating sign. As drones are able to see ultraviolet light the horns fluorescence acts as a visible sign to other drones. Figure 23 shows the *vestibulum* (meaning a large cavity) which consists of the large white hairy area that is surrounded by the two

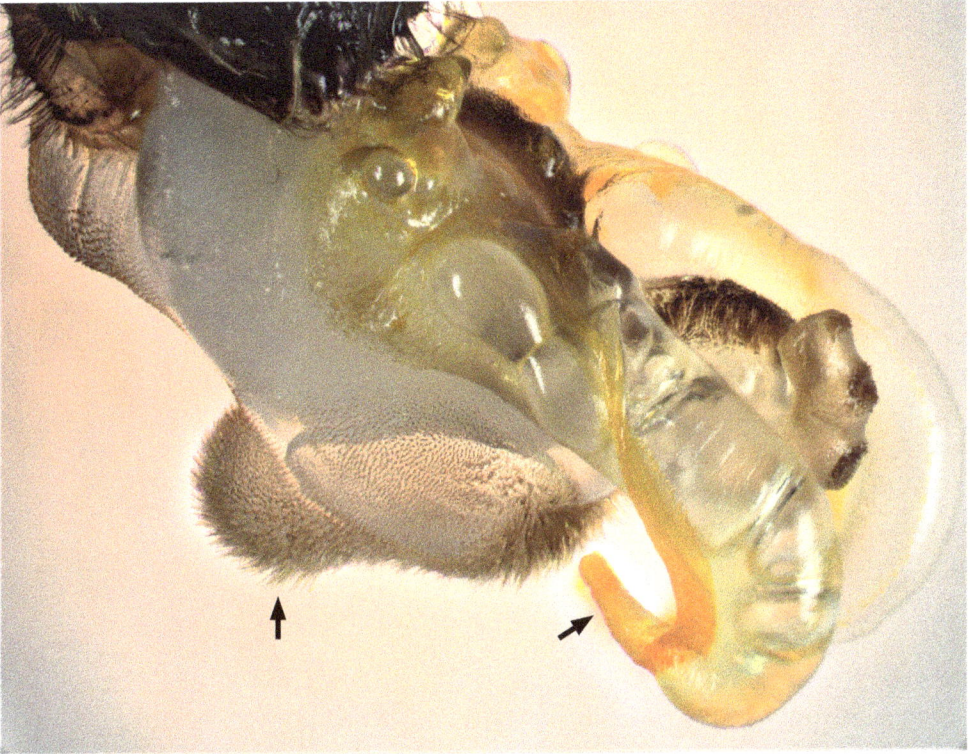

Figure 23: The drones vestibulum *and orange* cornua (horns).

cornua: yellow horns. Between these sits the *cervical* duct, a neck-like tube, through which stage two of the eversion takes place. Figure 24 shows a part eversion on an immature drone, the horns have no orange colouring unlike those of a mature drone.

The mechanics of eversion

The eversion of the *endophallus* results from the pressure of the *haemolymph*

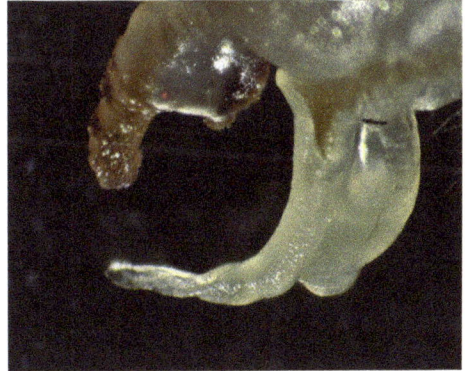

Figure 24: Part eversion of an immature drone lacking in orange membrane.

which is pushed backwards by the contraction of the strong *abdominal* muscles. First the *penis* valves and the claspers open (*laminae para-merales*), then comes the eversion of the *abdominal* wall with the basal part of the *endophallus* which carries the hairy triangular plate, and finally the two *bursal cornua* appear.

Inside the everted base (*vestibulum*) of the *endophallus*, the bulb with its *chitinized* plates has emerged (Drawing 3:3). Even in this early phase of eversion, the *lumen* (the internal wall) of the *endophallus* is connected with the outer surface between OB and OC (Drawing 3:1). **See key to abbreviations used at end of this chapter.**

Figure 25 shows the *bulb* in position behind the *cervix* (picture taken from a dead drone). The hairy area outside is where the mating sign from the previous drone is sometimes attached.

The connection between the *cervix* and the *bulb* consists only of a very small slit, and since its inner opening does not lie at the point of the *lumen* (inside space); there is generally no penetration of the *sperm* or *mucus* from inside to outside at this stage. *Sperm can* be ejaculated into the lumen of the *penis* bulb when the drone's abdomen is squeezed for *sperm* collection, but transfer does not occur at this stage in natural mating.

The most difficult part of the eversion, which requires the greatest force, is at the middle portion of the organ often referred to as the *cervix*. The sac of the *bulb* of the *endophallus* (Drawing 3:l, SB) and the ends of the *chitinized* plates enter the everting *cervix*, dilating it, then in consequence, rupturing the thin outer layer of its wall, producing one or more long fissures through the hairy areas on each side of it (Drawing 3: 6, FC).

Figure 26 shows the underside of everted *endophallus*. The transversal folds on the *cervical* duct and hairy fields are found on one side only; the three triangular areas (indicated by arrow) on top act as spreaders when eversion takes place,

Figure 25: **Bulb** *positioned behind* **cervix.**

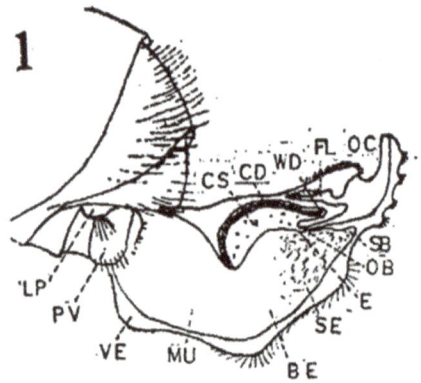

Drawing 3:1 in section. The continuous muscle contractions of the **mucus** *glands (MG); tear off the glandular epithelium, (cells) which is also pushed into the ejaculatory duct.*

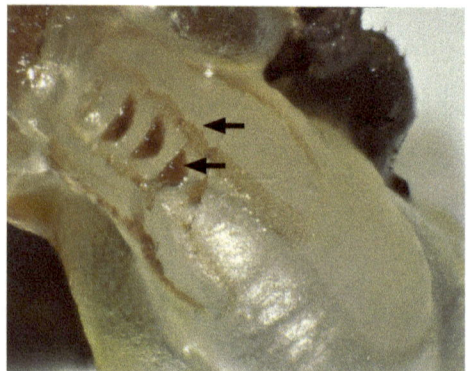

Figure 26: Underside of everted **endophallus.**

stopping adhesion to the wall linings and the longitudinal hairy band strips on either side.

At the same time the folds of the *endophallus* which come from the lateral *chitinized* plates are everted. These everted folds now form the terminal part of the everted *endophallus* (Drawing 3:2, CE). After this part of the *endophallus* has passed through the narrow part of the *cervix*, the whole *bulb* pushes very quickly rotates now faces backwards, towards the drones *abdomen* inside the everting *endophallus*. As a result the pointed ends of the *chitinized* plates quickly reach the end of the everted *endophallus*.

The ends of the lateral *chitinized* plates now protrude freely, externally, but the pointed ends of the dorsal *chitinized* plates are wrapped in two folds of the dorsal wall (Drawing 3:5, WD). The enveloping folds do not allow the plates to push themselves out of the *endophallus*, but give them a fulcrum about which they will turn, later on in the eversion.

After complete eversion of the middle portion, and the freeing of the pointed ends of the lateral plates (Drawing 3:2, CL) the dorsal wall of the sac of the *bulb*

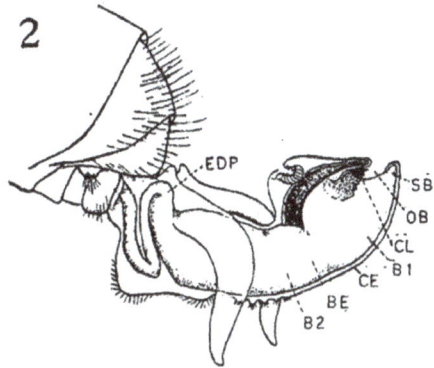

Drawing 3:2. The small tube has to be stretched open by the excessive blood pressure in order for the **bulb** to make its way forward, the **bulb** is 2mm in diameter and the tube has an internal diameter of about 0.5mm. See Drawing 3:4.

Figure 27: Partial second stage of inversion of **endophallus**.

everts. This causes a considerable widening of the slit-like aperture of the *endophallus* (Drawing 3:2, OB); as a result, *sperm* begins to flow out from inside the *bulb*, being pushed out by the *mucus*. (Mucus may be deposited in the mated queen to prevent the escape of *spermatozoa*.)

The *bulb* fills almost the whole *lumen* of the everted part of the *endophallus* (Drawing 3:2 BE). The thin-walled part of the *bulb* is everted more quickly and easily than the dorsal wall, with its *chitinized* plates, which meet considerable resistance.

So the sac of the *bulb* everts (Drawing 3:3, SBO, and the *chitinized* plates are almost vertical within the *endophallus* (Drawing 3.3). The *semen* and *mucus* in the secretory ducts, which are under pressure from the muscles and *haemolymph*, meet with no resistance and can easily spurt out.

After ejaculation the membranous part of the *bulb*, which has so far been much dilated (Drawing 3:3, B2), collapses (Drawing 3:4, B2).

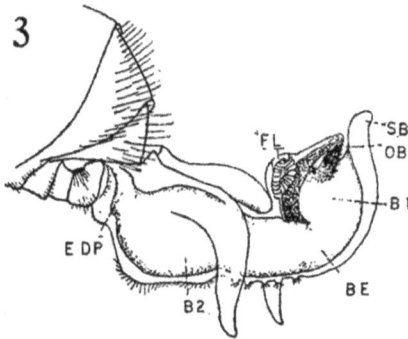

Drawing 3:3. It frequently happens that the two mucous *glands and* seminal vesicles *are now pushed into the empty space within the base of the endophallus (Drawing 3:4, MG, VS).*

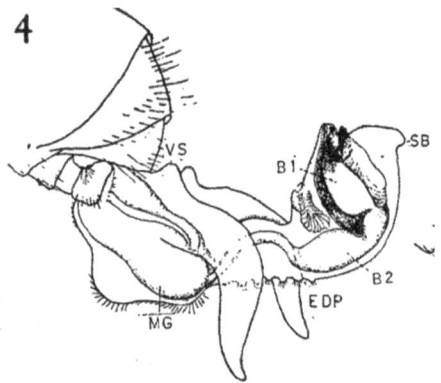

Drawing 3:4. Meanwhile, under the pressure of the haemolymph, the eversion of the endophallus continues and the chitinized plates of the bulb incline more. Because of this, the thick-walled part of the bulb is also everted, and the lateral plates and the membrane of what was the ventral side of the bulb, the bow of the bulb, now stand free (Drawing 3:6).

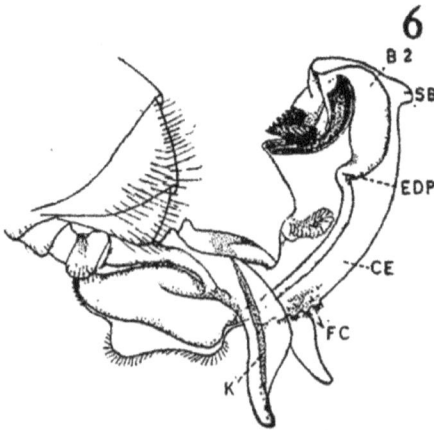

Drawing 3:6. After the eversion of the chitinized plates, the full eversion of the endophallus is completed quickly and comparatively easily. Because the ventral wall of the bulb everts more powerfully and more quickly than the dorsal wall, the membranous end of the bulb curves upwards and forwards, until the opening at the end of the endophallus is turned into a position within the everted plates. Here the remaining contents of the bulb flow out (Drawing 3:7).

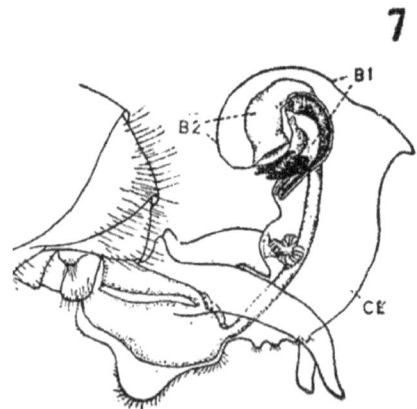

Drawing 3:7. In the final phase of eversion, however, the end of the endophallus (with the opening of the ejaculatory duct) is free and points forwards, towards the drone's head. This phase is well illustrated in (Drawing 3:2).

The *fimbriate* lobe is everted only at the finish. The pressure of the *haemolymph* and air in the everted *endophallus* can be so great that the thin wall at the end of the *bulb* bursts, and the *haemolymph* flows out.

About 30% of matings return with no mating sign. In the queen the *chitinized* plates of the mating sign reach as far as the *bursa copulatrix*. Indeed the points of the lateral plates project into it, and the dorsal plates press in under the folds of its lower edge. The broad ends of the *dorsal plates* lie just above the tip of the final *sternite* of the queen. The sting apparatus is pushed upwards and backwards by the mating sign, so that the sting, usually hidden, projects freely.

1. The ejaculation of the *sperm* does not occur when the *endophallus* is wholly everted, but when eversion has been carried out as far as the *bulb*, and the points of the lateral *chitinized* plates (now projecting free) open up a passage out of the *lumen* of the uneverted *bulb* (Drawing 3:2). The *sperm* only enters the *bulb* during eversion, and from there is flung out on the sudden production of an opening in the *bulb*.

Figure 28 shows a stripped *bulb* in the final position in the end pocket of the everted *endophallus*. *Mucus* and *semen* are at the end (see also Drawing 3:11).

2. If during the process of eversion the end of the *endophallus* meets with some resistance, the *bulb* cannot evert fully. Its inner lining (the *chitinized* plates together with a bow-like membrane referred to as the bow of the *bulb*) is pushed out of the *endophallus* uneverted.

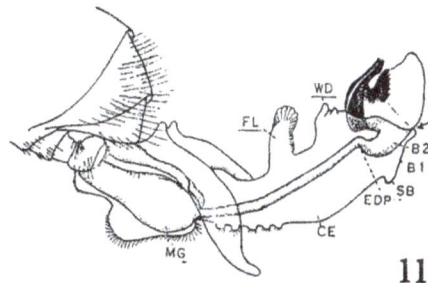

Drawing 3:11.

The separation from the *bulb* covering occurs along a layer of *chitin*, whose outermost stratum remains on the *endophallus*. This body, the *chitinized* plates with the bow of the bulb, filled with *mucus* and covered outside with the yellowish inner *chitin* layer of the *bulb* wall, forms the normal mating sign.

3. The *endophallus* is undamaged when it becomes detached from the mating sign. After the end of the eversion, the pressure within can result in its wall bursting, with an audible pop.

If the wall is torn earlier, its turgidity is lost, and the complete detachment of the *chitinized* plates is then much more difficult, and takes longer.

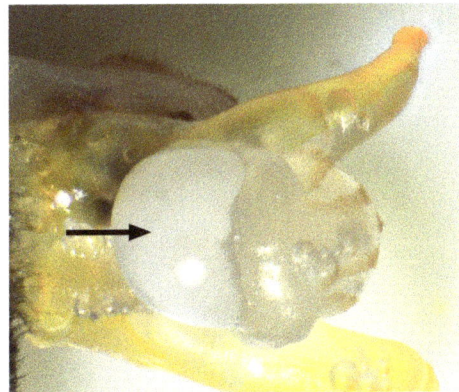

Figure 28: Stripped bulb in the final position showing mucus.

It is therefore not necessary to assume that the *endophallus* is broken or twisted off. The separation of the mating sign takes place, under suitable conditions, of its own accord and very quickly.

At the commencement of mating the base of the *endophallus* and the pneumophyses (small lobes) are the external plates surrounding the abdomen exit. They are the first organs to be everted. The points of the latter are first turned dorsally and outwards.

As the internal pressure increases, the sticky, orange yellow surface layer of the *pneumophyses* becomes smooth and is pushed to their lateral sides. As a result the *pneumophyses* bend downwards; their points become closer, probably embracing the queen's *abdomen*.

The *bulb*, which only becomes filled with *sperm* and *mucus* during the mating process, has meanwhile entered the wide cavity of the *vestibulum* (Drawing 3:1), and is now (under considerable pressure) pushed slowly through the everting *cervix*.

The partially everted *penis* is introduced into the actively opened sting chamber of the queen, the points of the *chitinized* plates now forming the end of the *endophallus* reaching into the *bursa copulatrix*. As eversion proceeds, the points of the lateral *chitinized* plates push out of the *endophallus* (Figure 30:2), and the sac of the *bulb* begins to evert (Drawing 3:3). As a result, an opening is formed which leads to the *lumen* of the *bulb*, and out of this opening the *sperm* is ejected under strong pressure. The *semen* is guided by the lateral *chitinized* plates into the *vagina* and from there past the depressed valve fold into the *oviduct*.

The viscous *mucus* which follows can only move slowly. It fills the *bursa copulatrix* with its side pockets, and the sting chamber, but does not normally extend as far as the *oviducts*. Since ejaculation of the *sperm* takes place from the uneverted *bulb* of the *endophallus*, there is no cogent argument for assuming that the position of the queen during mating must be above the drone. The further eversion of the *endophallus* encounters resistance in the narrow sting chamber.

The *endophallus* pushes against the queen's *abdomen*, and the *bulb* wall is everted under increasing pressure from within. Its contents, the *chitinized* plate's filled with *mucus* and the bow of the *bulb*, are pushed out of the wall uneverted. The separation takes place along the surface of a preformed layer of *chitin* in the *bulb* wall. This peeling off of the wall from the *chitinized* plates and the bow of the *bulb* begins at the tip and continues to the base (Drawing 3:12 –13).

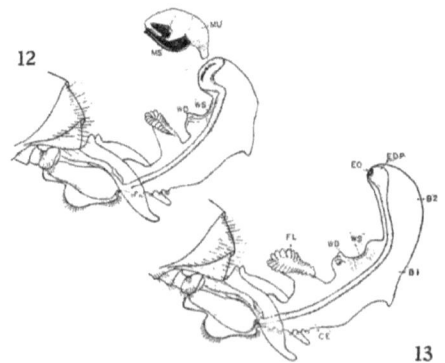

Drawing 3: Mating sign. 12: Separation of the mating sign from the endophallus. 13: Endophallus after the separation of the mating sign.

Figure 29: Full eversion: Compare its size to the abdominal size. See Drawing 8: Dade drawing plate 15.

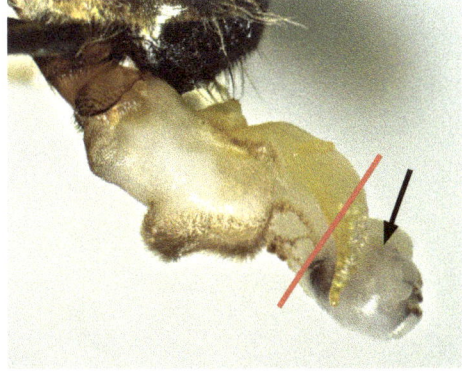

Figure 30: This area of the endophallus is left behind after mating; acting as a visual attraction to other drones, See Drawing 6: Dade drawing plate 14.

The now fully everted *endophallus* separates undamaged from the mating sign (Drawing 3:13), and with it the separation of the queen and drone is accomplished.

It is possible that the *fimbriate* lobe, a petal-like structure that inflates with *haemolymph* at its eversion helps to free the mating sign from the *endophallus*.

When the mating is finished the *endophallus* is fully everted, but it has lost its *chitinized* plates and the bow of the *bulb*. If at this instant the *endophallus* is under a strong pressure from within, it may burst at its tip. During mating, the flight of the pair is largely hindered, but the pair only falls to the ground if the queen and drone have been unable to part immediately, or if they have paired at a low height. The queen flies on after separation from the drone, and the dead drone falls to earth.

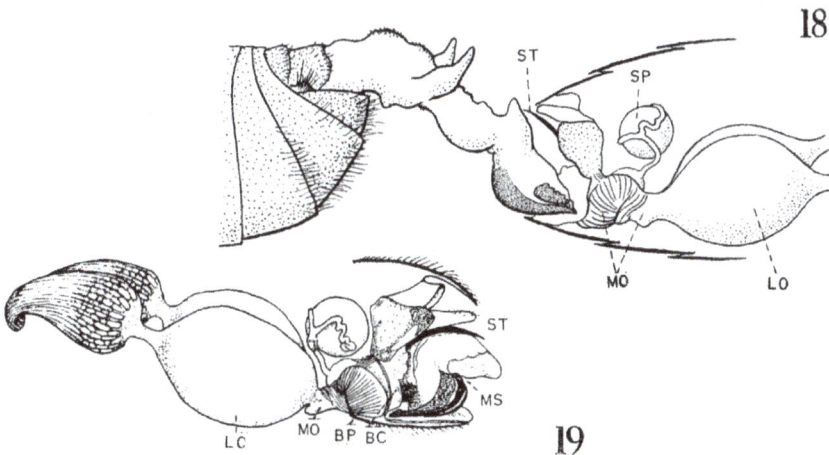

Drawing 3:18: Pair in copula: *in this preparation the great mass of* sperm *in the lateral* oviducts of the queen has been made visible. 19: *Position of the mating sign in the sting chamber of the queen.*

Drawings numbered 1-19 are reproduced by kind permission of Professor Jerzy Woyke 2016.

Key to abbreviations used in figures

BE   Bulb of endophallus; penis bulb

B1   The part of the bulb which is strengthened by the chitinized plates and the bow of the bulb

B2   Thin-walled part of the bulb

BC   Bursa copulatrix

BP   Bursal pouch

C   Pneumophyses of endophallus. Bursal cornua

CD   Dorsal chitinized plates

CE   Cervix of endophallus

CL   Lateral chitinized plates

CS   Connective substance; 'semi-transparent thickening of the cuticular intima

ED   Ejaculatory duct

E   Edge of connective substance partly concealing the posterior opening of the bulb of the endophallus

EDP   Posterior end of the ejaculatory duct

EO   opening of the ejaculatory duct; Gonopore

FC   Fissures in the outer layer of the cervix

FL   Fimbriate lobe

K   Sticky surface layer pushed back from the pneumophyses

LO   Lateral oviduct

LP   Lamina parameralis; upper claspers

M   Covering of the mating sign

MG   Mucous glands

MO   Median oviduct

MS   Mating sign

MU   Mucus

OB   Posterior opening of the bulb

OC   External opening of the cervix

PV   Penis valve; lower claspers

S   Bow of the bulb

SB   Sac of the bulb

SE   Semen

SP   Spermatheca

ST   Sting

T   Thread-like end of the mating sign.

VE   Vestibulum or base of the endophallus

VS   Seminal vesicle; vesicula seminalis

WD   Doming of the bulb wall, surrounding the pointed ends of the dorsal plates

WS   Stretching of the bulb wall for the chitinized plates

The seven stages of mating

In summary, seven characteristic mating stages may be determined during the process of natural mating.

At the beginning of natural mating, the drone becomes paralysed. However, the muscles in the *abdomen* continuously contract shrinking the *abdomen* till mating has ended and the pair have separated.

The drone which fails to remove the mating sign of its predecessor during the nuptial flight determines the age at which the queens start oviposition. If the drone is say 15th to 20th to mate, it means the queen will collect a lot of *semen* during the first nuptial flight. Such a queen starts oviposition 2 days later, at the age of approximately eight days. However, it may happen that already the second or the third drone fails to remove the sign of the predecessor. Such queen will collect less *semen* and often conduct more nuptial flights, delaying the start of oviposition later (mean; 13 days). If a queen must wait several days due to meteorological conditions, she will initiate oviposition at an even older age; up to 29 days old.

1. At the beginning of the mating, the drone brings the end of his *abdomen* closer to the queen's open sting chamber. The drone's *abdomen* muscles contract and the *endophallus* is everted into it.

During the first mating stage, the *endophallus* is only partly everted into the queen's sting chamber. The curved slender tip of the partly everted *endophallus* is inserted into the queen's vaginal orifice (Drawing 4:3, MS).

It is only the head and the *thorax* of the drone that becomes paralysed. The *abdomen* remains active. The *abdominal* muscles continue to contract during the whole mating process up to the separation of the pair.

This is controlled by a set of the *ganglion* cells that are found in several places in the bee's body.

In the same way, the sting that is torn out of the worker also contains a *ganglion* and this acts on the muscles to keep injecting *venom* into the victim. At the beginning, the *bulb* inside the partly everted *endophallus* in the sting chamber is empty.

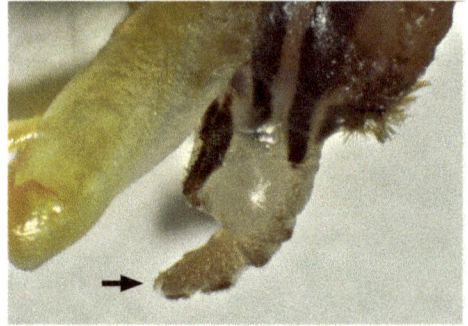

Figure 31: The small tube has a natural upward curve to aid in insertion into the queen's sting chamber and then the second stage of eversion takes place and ejaculation into her ovary ducts follows.

Despite the drone's head and *thorax* being paralysed, the muscles of the *seminal vesicles* (Drawing 4:1, SV) contract and push the *semen* into the *ejaculatory duct*. Next, the muscles of the *mucus glands* (MG) contract and push the mucus into the ejaculatory duct. The *mucus* pushes the *semen* forward and fills the *bulb* (Drawing 4:3).

Drawing 4:3. Mating stage (MS) 1. (Longitudinal section.) The slender tip of partly everted *endophallus* is inserted into queen's vagina. No *semen* is in the tip. Orifice of *cervix* OC is tightly closed

The *epithelium* pushes the *mucus* forward, and the *epithelium* also fills the posterior part of the *bulb* in the partly everted *endophallus* in queen sting chamber. No *sperm* is present in the tip of the *cervical* thin tube, (Drawing 4:3, MS).

The internal side walls of the *cervix* are provided with special strong hairs which keep the walls very tight together. The *endophallus* stops at the partly everted stage during eversion because of this structure. A slender curved tipped, thin tube appears at the end of the *cervix*. No opening is present in this tip. However, a line fissure is visible at the dorsal wall of the curved tip. The edges are tightly closed together.

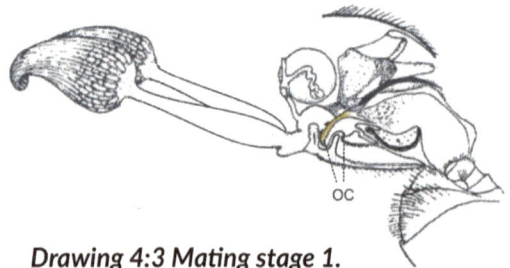

Drawing 4:3 Mating stage 1.

2. During natural mating, the eversion of the *endophallus* stops for a while after the partial eversion.

Higher pressure is required to reach the second mating stage. When the required pressure is reached, the large *bulb* filled with *mucus* and *sperm* is pushed by force through

Drawing 4:4 Mating stage 2. The bulb passed through the cervix. The semen is injected into the oviducts.

Figure 32: Mating stage 2 Simulation, the earliest eversion stage, the semen can be injected.

the *cervix* and the thin tip at the end of the *cervix*. Only after the *bulb* passes through the *cervix* and the thin tipped tube, the tips of the longitudinal *chitin* plates and the orifice of the *bulb* appear at the end of the so far everted *endophallus* (Drawing 4:4 and Drawing 32:15b, MS).

Only now, the *semen* may come out of the *bulb* (Figure: 32). During natural mating the *sperm* is injected by force out of the *bulb* at the end of the *endophallus*, into the common *oviduct* of the queen.

(At this eversion stage the *sperm* may be collected also for instrumental insemination, when manual pressure is applied to the *abdomen*.)

3. After the *sperm* is injected into the oviducts, the third characteristic mating stage MS 3 occurs.

The *chitin* plates together with the *mucus* and the *epithelium* slide out of the *endophallus*, and the mating sign is created (Drawing 4:6 -7). The mating sign remains in the sting chamber of the queen. The *endophallus* becomes fully everted.

4. However, it is still attached to the end of the mating sign (Drawing 4:8); presenting the fourth stage MS 4 of the mating process. The pressure inside the *endo-*

Drawing 4; 6 Mating stage 3. The mating sign sliding out of the bulb. Mating sign of the predecessor removed.

Figure 33:7 MS3. Simulation, creating the mating sign, the queens open chamber on left side.

phallus is still so high that it squeezes the ejaculatory duct inside the *endophallus*.

As a result, the *epithelium*, which was torn from the *mucus glands*, and now is present inside the *ejaculatory duct*, is pressed out of the *endophallus* (Drawing 4:9, TT). The *epithelium* appears as a whitish thread at the end of the mating sign.

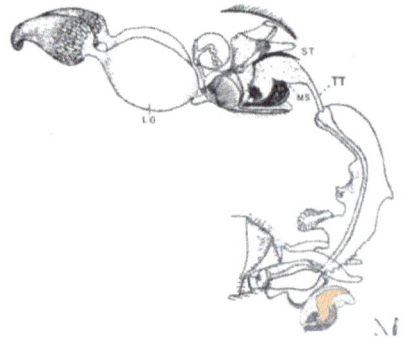

Drawing 4:8 Mating stage 4.

5. The *endophallus* is still hanging at the end of this thread (Drawing 4:9, MS 5).

6. The thread is not able to support the drone, and thus it breaks a short distance from the end of the mating sign (Drawing 4:10, MS 6). Some queens return to the hive with this thread.

Therefore, the end of the mating sign looks blunt. The *endophallus* has neither nerves nor muscles. Also during full eversion, the mucus glands and the *seminal vesicles* are pushed out of the abdomen into the *vestibulum* of the *endophallus* (Drawing 4, 6.-8-9).

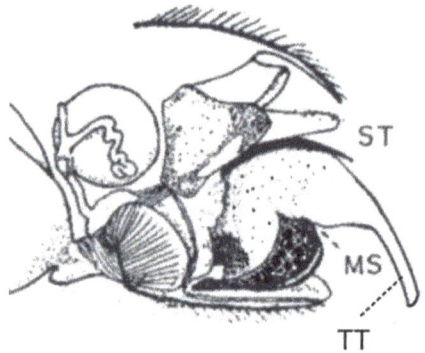

Drawing 4:9 Mating stage 5.

No active role needs to be played by the queen, nor is it possible. The contraction of a queen's sting chamber is not able to initiate full eversion of a partly everted *endophallus*.

When a drone is manually provoked for partial eversion of the *endophallus*, no external manipulation will stimulate full eversion of it. Only the pressure of the abdomen will result in full eversion of the partly everted *endophallus*.

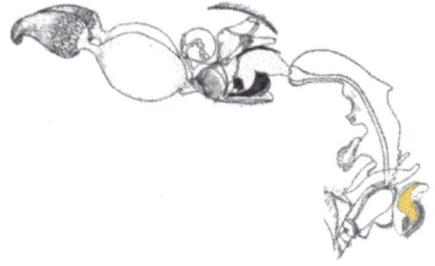

Drawing 4:10 Mating stage 6.

Sequential mating occurs when the mating sign from the previous drone is inside the queen's sting chamber. Despite the chamber being filled, it is easy to open the last abdominal *sclerite*. The partly everted *endophallus* of the next drone can be introduced between the *sclerite* and the mating sign.

The *cornua* of the next drone embrace the mating sign of the previous drone. The cornua may serve as *parameres* (like genital claspers), which in other insects are more developed, fixing the pair in *copula* and directing the slender tip of the *endophallus* into the queen's *vagina*.

7. The *cornua* bend down because the orange sticky membrane bursts at the dorsal wall of the *cornua* and the membranes slide down toward the ventral wall of the *cornua*. The orange sticky membranes stick to both sides of the mating sign of the previous drone inside the sting chamber of the queen.

Progressive eversion of the *endophallus* results in a pulling of the mating sign of the previous drone out of the queen's sting chamber. The sign becomes attached to the basal, hairy, *vestibulum* field of the *endophallus* of the next mating drone (Drawing 4: 6- 8-9).

Thus, the removed mating sign originates from two drones. The *chitin* plates and the mucus originate from the previous drone, but the orange membranes originate from the next mating drone, which removed that sign (Drawing 4, 6-8-9). The mating sign left in the queen's sting chamber by the last drone is not covered by orange membranes (Drawing 4: 8 -9- 10). The *cornua* of the *endophallus* of the drone, which recently mated with the queen, are also without those orange membranes. Instead, the orange membranes remain on the removed mating sign.

Drawing 4: MS 7. Mating sign covered with orange membranes. The thin thread curved (TTC) on the mating sign (MS).

In any subsequent mating, the drone mates with the queen and his *cornua* embrace the mating sign (Drawing 4:11). However, the drone that fails to remove the mating sign of the predecessor means that the mating sign in the sting chamber of the queen returning to the colony, could originate from two drones.

The *chitin* plates, the *mucus*, and the *epithelium* originate from the last drone, but the orange membranes covering the mating sign are from the next drone, which failed to remove the predecessor's sign. Some queens return to the colony with two pairs of orange membranes.

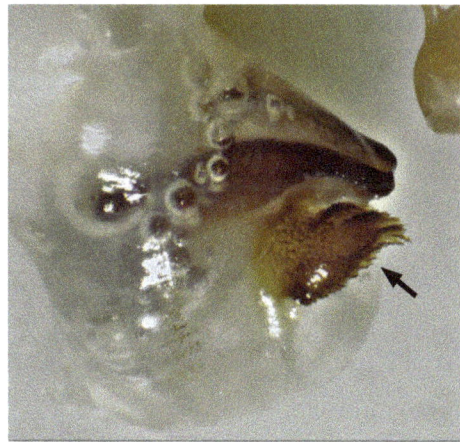

Figure 34: Everted bulb: Showing lateral plates which form part of the mating sign. See Drawing 4:9.

There are also instances of queens, which return with an additional incomplete mating sign. This phenomenon indicates that after the drone failed to remove the predecessor's mating sign, the next drone, which tried to mate with the queen, also failed to remove the mating sign.

The above example shows that the mating sign in the sting chamber of a queen returning to the colony, originates from two or more drones.

The difference between provoked eversion and natural eversion is that the chitin plates do not slide out of the *endophallus*, because the backpressure of the queen's sting chamber is missing. However, when the free eversion is artificially hindered, the chitin plates also slide out (Drawing 4:7). Of course, during natural mating, the *semen* is injected into the queen's *oviducts*, while during provoked full eversion semen and mucus disperse on different parts of the fully everted *endophallus*.

The Mating Process

Queens and drones mate in the air during the mating flights. The duration of the mating flights varies between 15-30 minutes for the queen.

The mating act takes less than two seconds from the drone grasping to releasing the queen, meaning that mating with 1, 8, or 15 drones would take 2, 16 and 30 seconds, plus drone positioning time. Thus, mating with several drones, one after another, can be completed in less than one minute.

Once a queen mates with one drone, a thin whitish thread is hanging at the end of the mating sign (Drawing 4:10, TT). However, in almost all queens returning to the colony, the thread is bent forward on the mating sign, making the mating sign look blunt.

The next drone mating with the queen pushes and bends the thin thread forward at the end of the mating sign of the predecessor. When an unsuccessful drone fails to remove the mating sign of his predecessor, the queen returns to the colony with a blunt mating sign. Thus the failure to removing the mating sign from the queen's sting chamber determines the cessation of the nuptial flight.

The nuptial flight is terminated because the last drone trying to mate with the queen is not able to remove his predecessor's mating sign from the sting chamber of the queen.

It is important to note, that all successfully mated queens return to the colony with the mating sign covered by the orange membranes. This indicates that a drone tried to mate with the queen. His *cornua* had embraced the mating sign of the predecessor. The orange membranes are stuck to the mating sign, but the drone failed to remove the mating sign of the predecessor.

It can happen that the orange membranes covering the mating sign are not clearly visible. Instead they are covered by mucus, and some *semen* may be found somewhere on the surface of the sign, both the elements originate from the next drone, which failed to remove the mating sign.

The whole above explanation shows that the nuptial flight terminates because a drone fails to remove the mating sign. Other drones, which tried to mate with the queen, also failed to remove the sign.

There are varying degrees of difficulty needed to dislodge the mating sign from queen's sting chamber. Also, the ability to remove the sign is not perfectly developed in all drones.

The behaviour of different queens, and the anatomy of the sting chamber, probably also vary.

The above information shows that the queen feels the quantity of *semen* she collects in her *oviducts* during the nuptial flight. She recognizes when she has not gathered sufficient quantities of *semen*. She then wants to mate with more drones. One drone can more than fill the *spermatheca*, but the capacity of the *oviducts* and *bursa copulatrix* is much higher than the *spermatheca* to enable mixing of *semen* from different mates. This implies or requires some mixing procedure prior to migration to the *spermatheca*. Thus the queen must have some kneading mechanism - pulsating, rippling, and stirring to create the mixture.

Drone *sperm*

Drone *sperm* have very long coiled tails compared to other social insects about 250 microns long. There are about 10 million *spermatozoa* in the drone's seminal vesicles.

It is thought as *sperm* use their tails to swim to the *sperm* duct where they form *sperm* clumps, any remaining dead or damaged *sperm* are likely to be left behind, due to low mobility.

Clumping indicates that at this stage *sperm* from different drones is non-competitive. *Sperm* are also believed not to be competitive within the *spermatheca* once mixed.

Competition between *sperm* for fertilisation of an egg mainly happens when females mate with many partners, as with honey bees. This is called polyandry.

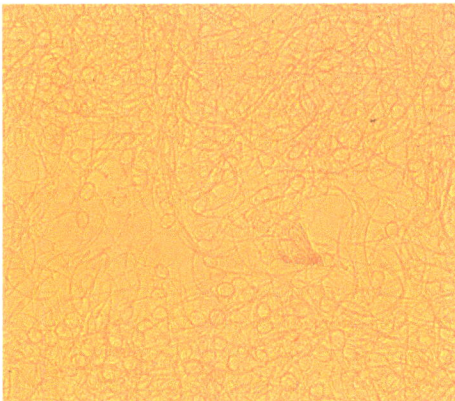

Figure 35: Tangled sperm (Unstained 400 x magnification) see Figure 3: Dade drawing plate 30.

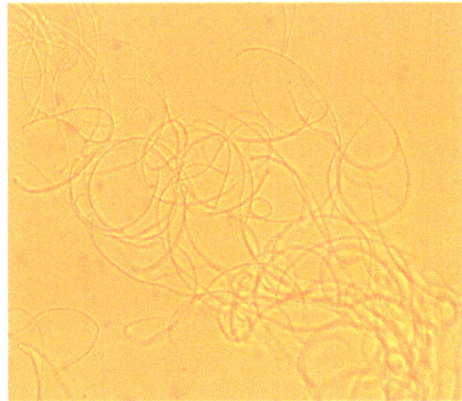

Figure 36 A: Sperm unstained (1000 x magnification).

As queens are believed to release between 2 and 25 *sperm* when fertilising an egg, it is thought that it is at this stage *sperm* are using their long tails to swim competitively, ensuring only the fittest fertilises the egg.

Seminal fluid, a potential problem

When females mate with multiple males, seminal fluid components can become agents of sexual selection and harm rival ejaculates. This does not seem to apply to honey bees at this stage of mating (*semi-nal fluid* is a complex mixture of proteins and metabolites with multiple functions, it keeps *sperm* alive and motile, protects against pathogens, and regulates *sperm* capacitation, the final maturation step that enables *sperm* to fertilise eggs).

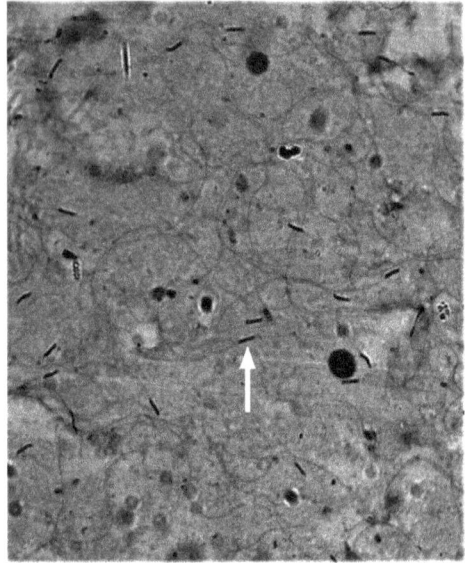

Figure 36 B: eq sperm stained, showing heads and tails. (1000 x magnification).

It has been found that *seminal fluid* can manipulate the queens physiology by causing deteriorating vision. This reduces her likelihood of leaving the hive after she has mated the first time, however if she is to mate again due to insignificant sperm, she will try to counteract these effects by leaving for mating flights earlier in the day, thereby increasing offspring genetic diversity and the success of their colonies.

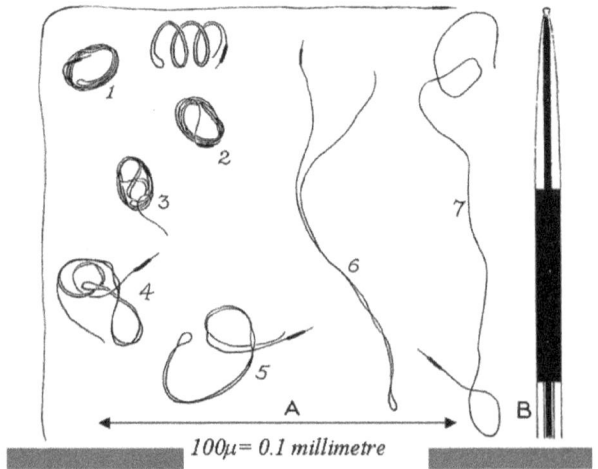

Drawing 5: A. spermatozoa, as they appear in the stained smear: 1, 2, coiled, inactive: 3 to 7, stages in uncoiling.

The total length of a spermatozoon is about 0.25 mm; the head is about 10 microns long by 0.5 microns in diameter. B. Structure of the head and part of the tail. (A, drawn from a smear; B, simplified after Rothschild.). Bar represents 100 microns.

100μ= 0.1 millimetre

Taken from Dade's book: Plate 30: Anatomy and Dissection of the Honeybee. Kind permission to use them has been obtained from copyright holder IBRA.

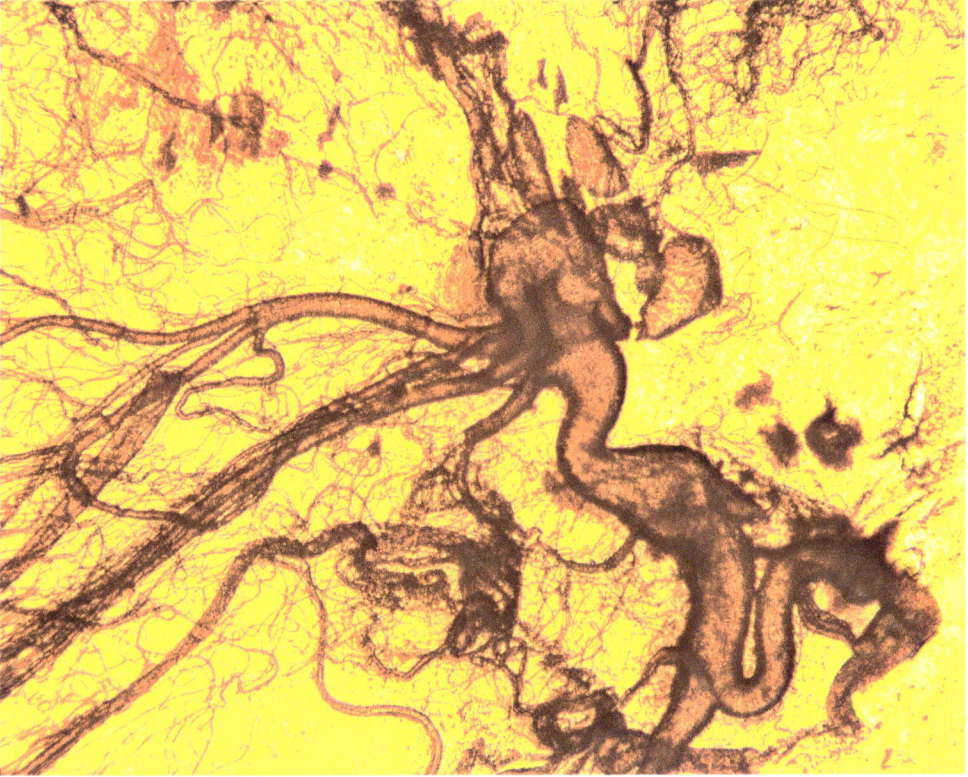

Figure 37: **Tracheal** *system. This shows the vast oxygen providing network that supports the outside of the testes aiding in* sperm **growth. (400 x** *magnification) (A similar system surrounds the queen's* **spermatheca,** *her* sperm *storage organ)*

The *sperms* develop and mature in both *testes* of the drones and are composed from about 200 tubular *testicoles*. The primary *sperm* cells, the *spermatogonia* divide many times increasing their number up to 10 millions, next they grow into large *spermatocytes*. Later they become elongated as *spermatids* and finally, they are transformed into *spermatozoa*, with the head at one side and the tail at the other side. Inside the head is the nucleus, and inside the tail are structures enabling the movements of the *spermatozoa*.

The final development of *spermatozoa* is accomplished four days prior to emergence of the drones from the comb cells. Three days after the emergence of the drones from the cells, the *spermatozoa* go down into *seminal vesicles* via the *vasa differentia*. At this time, the *spermatozoa* are still not mature enough to fertilize the eggs. The internal walls of the *seminal vesicles* are lined with *epithelial* cells. The spermatozoa direct their heads toward the *epithelial* cells and use the secretions of the *epithelial* cells to become mature. The *spermatozoa* remain in seminal vesicles during the whole life of the drones.

It is interesting to know, that the drones produce all the *spermatozoa* only once, in their development stages, before emerging from the comb cells. The *spermatozoa* of *diploid* drones reared by Woyke have the same structure as those of *haploid* ones. However *diploid spermatozoa* are larger than the *haploid* ones and the diameter of the tail of *diploid spermatozoa* is 115% of that of the *haploid* ones.

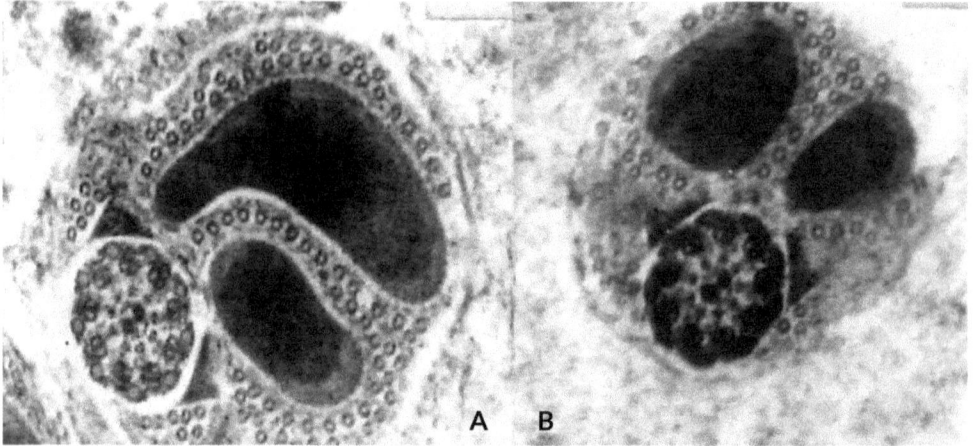

Fig 36C: Cross section through tail of spermatozoon (x 66 000 magnification) A. Diploid drone B. Haploid drone

Drone production and development

The strain of the bee will influence the number of drones laid and when. The Italian queens, the most popular strain in Britain, produce drones early in the season.

Drones are reared in larger cylindrical cells (6.3 mm diameter) than workers (5.3mm) on a 9-13° angle from the horizontal and have raised domed cappings, the whole within a supporting hexagonal comb structure.

They are mainly found on the outside of the brood nest, often at the bottom of the frame. Overall, typically about 15% of the brood nest comprises drone cells. Unfortu-nately for the drones, *Varroa* mites prefer drone comb in which to reproduce; they have more room to develop and more importantly, extra time to reach matu-rity, 24 days compared to 21 days for workers.

Drone rearing starts at the beginning of the foraging season when nectar and pollen become plentiful.

Drone eggs are therefore laid from late spring to late summer and peak drone population occurs during swarm season. All colonies in the same geographic area

Figure 38: Lateral view: Late stage of drone pupae, a day or so away from exiting the cell. 7th Instar.

are usually synchronised to start drone rearing at the same time but drone production tends to start earlier in colonies with large worker numbers.

Colonies with larger food reserves produce more drones. In contrast, drone comb is not constructed in colonies with less than 4,000 workers and drones are not produced at all in small colonies or those with a high disease burden or when environmental conditions remain persistently unfavourable.

To ensure a sufficiently large drone population has been established in the locality prior to the first virgin queen mating flights, drone rearing starts about three to four weeks before queen rearing. However, other than drone rearing preceding swarming, the two activities are not closely correlated within individual colonies.

Whilst swarming colonies tend to produce more drones, the presence of drones per se is not predictive of swarming, although their absence can be a contra-indicator. Successful drone rearing is draining of colony resources and rests upon the maintenance of exacting nutritional and brood nest conditions.

The presence of drone brood slows down the rate of further drone egg-laying by the queen. Furthermore, she will not lay in drone cells in very poor weather.

Drone eggs and *larvae* are cannibalised by workers whenever weather conditions or colony nutritional reserves are very poor. No eggs at all are laid in drone cells from late summer onwards.

In the late summer from the end of July to the beginning of August drones are no longer fed by workers and they are evicted from the nest to die. This eviction is delayed if colonies receive autumn feed and it does not happen at all in queenless colonies or failing queen colonies.

Although about 5,000 eggs will be laid in drone cells annually, only about 2,000 drones will develop into sexually mature bees, a 40 % success rate.

As previously stated drones take the longest time to develop of the 3 castes, 24 days. A queen takes 16 days, workers 21 days.

One reason for the time difference is because the drone's body is given over to the development of the testes. Primordial germ cells develop in drone *larvae* the moment they hatch from the egg. *Spermatogenesis* takes place rapidly, resulting in the production of spherical *sperm* cells that have a head but lack tails.

All *sperm* cells have been synthesized by the sixth day of *larval* development. They mature, but they don't multiply

Figure 39: Drone raised capping.

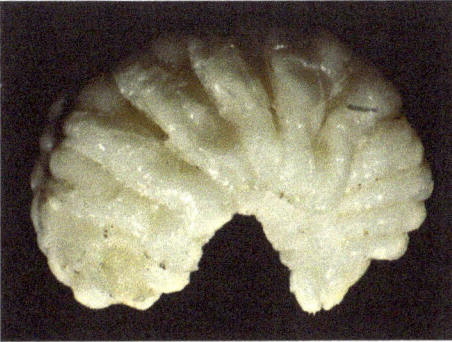

Figure 40: Drone pupae
about 7-8 days old.

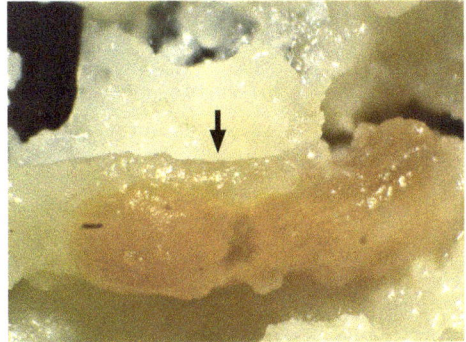

Figure 41: Same drone pupae as
drawing 6, showing testis.

during development and continue to
mature into functioning *sperm* for about
12 to 14 days after emerging from the
comb. At this time the drone becomes
sexually mature. *Sperm* degrades during
the lifetime of the drone so peak fertility
is not maintained.

Sperm will then be stored in the drone's
seminal vesicle ready for mating and the
testes will atrophy.

Drone *larvae* need considerably more
food than worker *larvae* and receive
considerable quantities of brood jelly
which contains a wider range of proteins. Older *larvae* receive pollen and
nectar in their diet.

Workers are more attentive to the
thermoregulation of drone brood than
worker brood.

It is also at this time that the *sperm* cells
transform into *sperm*atozoa with heads
and tails.

Whilst initial development of the *endophallus* also starts during pupation and
the *sclerotized* plates are completely synthesized in this developmental phase, a

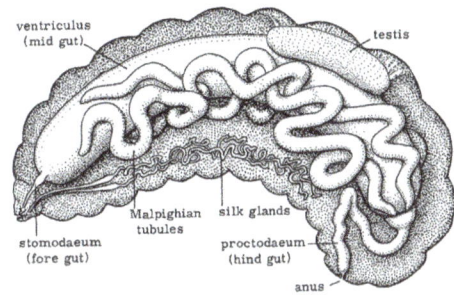

Drawing 6: Plate 19: Dissection of
larva from lateral aspect. Taken from
Dade's book: Anatomy and Dissection
of the Honeybee. Kind permission
to use them has been obtained
from copyright holder IBRA.

Figure 42: Drone eyes and flight
muscles develop during pupation.

considerable amount of the reproductive apparatus is only made post-eclosion, (after
the drones emerge from their sealed cell for the first time, day 24 of their lives.)

Figure 43: Some drones emerging from cells, note raised drone capping on the right.

Sexually immature drones

For the first eight days post emerging, young drones are fed a combination of brood food, honey and pollen by worker bees and they remain in the warmth of the brood area whilst undergoing their second phase of maturation, which is both physical and sexual.

It is only during this time that much of the reproductive apparatus first develops. *Sperm* only starts to migrate to the *seminal vesicles* on day 3 post emerging.

When fully mature, the *endophallus* and related glands are tightly packed within the *abdomen* to conserve space.

Six to eight days post emerging, drones start making their first cleansing flights which last about 2 and 3 minutes. At about this time, they move to the periphery of the hive and start to feed themselves.

Between days 12 and 15, nearly all the *sperm* has moved to the *seminal vesicles* where it attaches to a single layer of glandular cells on the inner surface to remain viable.

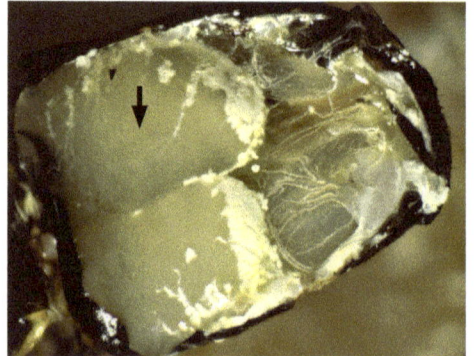

Figure 44: Sexually not fully mature testes: large area on the top side, of an immature drone just released from its cell at 24 days old: Note their size at this stage. See Drawing 7: Dade drawing plate 14.

Within the nest, drones are largely inactive but they will groom themselves and continue to beg for food. The high sugar diet is needed to power their regular afternoon mating flights on a warm day.

Drones can considerably differ in size depending on nutritional and environmental factors. There is evidence that having a larger body mass when drones hatch out of their cells, indicates that they will successfully reach sexual maturity and will be amongst the drones who attend the congregation areas.

When *haploid larvae* are raised in worker cells, which are smaller than drone cells, the resulting adult males are called "dwarf" or stunted but are still fertile drones despite their smaller size. Such drones produce fewer total *spermatozoa* than larger ones.

Drones take progressively longer orientation flights between days 10 and 13. At about two weeks post-eclosion, they start undertaking 25-minute flights to drone congregation areas where they look for virgin queen bees in need of mating. They are restricted by the amount of honey they can store in their *stomach*, so must return for food before leaving again, repeating this process for an average of 4 times per afternoon.

The average life span of a drone bee is 3 to 5 weeks post emerging, although it is possible for some drones to live up to 50 days. When a colony swarms then a very small proportion of its drone population will accompany them.

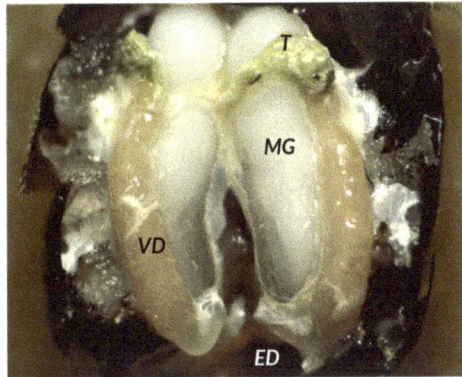

Figure 45: **Testes: T, small green shrivelled area on the top side. Vas deferens: VD** (*meaning, a conveying vessel*). **Mucus gland: MG, of a mature drone. Ejaculatory duct: ED: See Drawing 8: Dade drawing plate 15.**

Drone colour

The colour of the drone is dictated by the queen's parents, the drone's grandparents. If you have both a dark native and a yellow Italian drone in the same hive, and not visiting from other hives, the chances are that the queen's mother mated with a drone from a different race than her own.

The role of the drone in the hive

Honeybees are insects well known for their ability of social thermoregulation. This essential property allowed the originally tropical insect to expand its range into colder temperate environments and to survive there as a whole colony.

Nest temperature is maintained within a temperature range of 32–36 °C when brood rearing. For optimal development it is essential that the brood is maintained within this temperature range. While eggs and larvae in open brood cells can tolerate lower temperatures for some time, the pupae in capped brood cells are especially sensitive. Incubation of the brood has, for the greater part, to be accomplished by the worker bees which react to brood comb temperature rather than to air temperature.

In preparation for flight and during takeoff and landing drones always show a high thermogenic capacity with higher *thorax* temperatures than workers.

With their great body mass they are well positioned for contributing to the thermogenic needs of the colony. In young worker bees the ability of active heat production by means of the thoracic muscles develops in their first two days of life.

Adult drones exhibit a much higher metabolic rate than juveniles, which indicates that endothermy of juvenile drones is also not completely developed after emergence. This suggests that older drones contribute more to colonial heat production.

It has been shown that drones mainly react to high thermal stress caused by low ambient temperature, and in reaction to their local ambient temperature, they vibrate their flight muscles to help heat the hive in general rather than heat sealed comb.

Field expeditions and mating

The average queen mating flight normally last 5-30 minutes at distances up to several kilometres from the hive. Mature drones' mating flights have been measured at between 25 to 57 minutes from the hive.

Drones usually mate on a sunny day when temperatures are above 18°C and wind speeds are less than 18 km an hour and between 1400 and 1700 hours. They make 3-4 flights at heights between 7 and 17 metres, depending on wind speed, giving a total flying time of up to 3 hours per day.

Drones have been seen resting on plants in between flights. The end of mating flights are strongly affected by changes in ambient light. They congregate in specific areas waiting for virgin queens to arrive and these areas appear to be the in the same place every year.

These congregation areas attract a large number of drones from many hives thus helping to prevent inbreeding. Drones release a *pheromone* from their mandibular glands which help attract the queen and other drones ensuring a good genetic mix. The drones locate the queen first by sex pheromones released by the queen, then by sight at about 50 metres distance.

They then follow her in a cloud like group and once she has mated with the first drone they form a comet like pattern behind her because of the mating signal left outside of the queen's sting chamber, which acts as an attractant to them.

Drone congregation areas (DCAs)

It is widely evidenced that certain parameters must be obtained for successful mating to take place. Bee scientists, however, have produced differing results in describing congregation sites where drones gather and virgin queens attend for mating to take place, or even agreeing whether DCAs actually, or at least uniformly, exist.

The country parson, Gilbert White, first noted the phenomenon in Hampshire, England, in his 1792 diary, a site still reputed to be active.

Where DCAs exist, drones and virgin queens fly some distance to reach these specific sites to pair in flight, generally found to take place some 10–30 metres aloft.

In order for drones to leave the colony several parameters have to be met:

- Timing: after mid day when the sun is high, temperatures warmest and therefore before ~1700 hours.

- Weather: dry and cloud cover less than 7/8th across the sky.

- Temperature: above ~190C (this may vary across races and latitudes).

- Wind speed Force 3 or less (small twigs and leaves will be seen moving and flags lightly flutter).

Researching the field between 1948-1981, Professor Friedrich Ruttner and his brother Hans, succeeded by Prof. Dr Nicholas and Dr Gudrun Koeniger, have illuminated our knowledge of drone congregations, especially in the context of an Austrian alpine valley. However, researchers in the UK (Butler et al), and even the same German researchers operating in the flatter landscape of north Germany, have been unable to confirm the same concentrations in other landscapes.

David Cramp in his 2018 MSc thesis seeking and monitoring DCAs using an UAV, has experienced similarly disappointing results in New Zealand. The correlation he might have expected between various topographical features, slope, aspect, landscape features or usage were either not found or only weakly correlated.

Thus, while factors have been identified which are both susceptible to human interpretation and may have indicative value as to where a DCA might be found, there is no universal prescription. It may even be the case that drones are using senses we have not fully identified such as magnetosensitivity, polarisation of light or thermal currents. Despite the limitations of humans, drones' can turn up in a DCA within 10 minutes of importation into a locality. What is not obvious to humans is apparently perfectly obvious to bees. They may be aided by a drone pheromone produced in the mandibular glands. How the queen finds and then chooses from several local sites the mating area she will attend remains unexplained.

Traditional methods to obtain DCA information have included helium balloons lifting a net into which drones will fly using a tethered virgin queen or synthetic lure as attractant, and X band radar. Human operated drones, or lures revolving round a central spindle with a camera attached have captured some interesting footage of the mating act and behaviour proximate to the lens, but do not show the mass of drones or the comet that forms behind the queen when mating first takes place.

A simple method to ascertain drone activity is to present a baited lure with a queen or 9-oxo-2-decenoic acid (9-ODA) on a pole 4-6 metres in the air in an area where a drone congregation area may be anticipated. If they are there or nearby, drones will be attracted to the lure. Greater density may be found at higher levels, justifying the use of helium at that site. Prospecting the area will give an indication of population, size and extent and help to distinguish a drone flyway from a fully established mating site. Drones frequently travel along specific flight paths such as hedges to reach their DCA destination.

Drones have been trapped in DCAs 11 kilometres from their hives, but this is exceptional. The bulk of the drone population will attend a DCA within 5kms and often as little as a few hundred metres. A full stomach of honey will fuel flight for around 30 minutes. Drones may be allowed to visit other colonies to refuel as they are incapable of feeding on flower nectar. A further reason why drones may on occasion travel a considerable distance to mate is to distribute their DNA throughout a larger population, aided by multiple mating of queens. The queen seems to restrict herself to about 7 kilometres of flight.

Obviously it will help limit in-breeding if drones and virgin queens from the same colony do not mate. This can be prevented if they attend different mating sites. The DCA studied over years in Austria clearly demonstrated that drones attended near DCA sites and queens further. It further established a typical presence of ~10,000 drones from 240 colonies. The difficulty in exactly replicating these findings in other landscapes appears to indicate this is not necessarily true of all queens or DCAs.

Figure 46: Drones being attracted in the southern UK in
mid July to a lure that is baited with 9-ODA

While this would amount to an average of only 42 drones per colony at the DCA, in practice not all colonies provide an equal presence at every local DCA. At an area population level, to maintain its genetic representation each colony will have to provide around 6.5 matings per season (calculations in footnote). DNA evidence shows that within this average some colonies achieve over-representation in the gene pool of their local population.

Once the drones gather they seem to prefer an elliptical mating area; waiting for the virgin queen to arrive, and once a queen flies into this area they show intense interest in her. However once she leaves this area they tend to ignore her and stay confined ready for another visiting queen.

Drones' flight is more limited in range than that of workers, as they carry only relatively small amounts of honey as well as small glycogen reserves, and they are not providing themselves with nectar in the field as workers do. It is also important for each drone to be in peak condition in an attempt to gain a competitive mating advantage. Therefore, successful colonies require a difficult combination of quantity and quality in its drone pool.

If we also assume that in the absence of beekeeper intervention or significant disease attrition every geographic area naturally hosts an optimal honeybee population spread across numerous colonies, then we can predict that on average, as many colonies will fail each year as there are surviving swarms (the majority are said to fail to reach their second season).

As it is not obvious which colonies will succeed or fail at the time of queen mating and as the opportunity for any drone to mate is so slender, every drone will vigorously pursue every queen bee in need of mating that enters a drone congregation area.

It has been noted that drones of different races tend to fly at different heights, possibly to stop cross mating. As drone quality and fitness is of the utmost importance to the colony for the dispersal of its genetic legacy, considerable time and resources are invested in the development of the best possible specimens.

Footnote

Looked at across the population year, if 240 colonies each produce 1,000 drones per year on average, the season's drone population will have been 240,000. If 240 queens are replaced bi-annually (240/2 = 120) and queens are mated 12 times on average, then the chances of a drone achieving mating will be 0.60% or ~1:167.

Drones' reproductive organs

This section is intended to clarify and refresh the information given above.

The drones' reproductive system is made up of two *testes*; once they are fully formed the *testicular* tubes inside them is where *spermatozoa* are produced.

As the drones mature, *testes* lose size until they get reduced to 1/3 of their original size (pre-birth). They start developing in the *larval* stage and continue into the *pupae* stage where most of their development is complete before hatching out of the cell.

Two *vasa deferentia*, (meaning vessels that convey) connect the *testes* and the *seminal vessels*. These help convey and store mature *sperm* into the storage organ.

The *seminal vessels* nurture and produce nutrient for the *sperms*. Next are two *mucus glands* which produce a substance that solidifies in contact with air and water, but not with *seminal* secretions. It is possible they act as a plug inside the queen's sting chamber to stop the *sperm* from falling out after ejaculation.

The ejaculatory duct conveys the *sperm* to the bulbous part and copulatory organ; the copulatory organ in the resting state is uneverted.

There are two stages normally to inversion; the first is when the orange horns (the *cornua*) and the duct are inverted by low pressure. The duct, which is curved upwards, is then inserted inside the queen's *bursa copulatrix* then when the second stage inversion takes place.

Ejaculation then occurs with some force; the *endophallus* is detached from the drone's body as he flips backwards and the drone falls to the ground and dies as a result of mating.

Drones' *abdominal* muscles are highly developed. This is important from the physiological point of view, so that at the moment of copulation the *endophallus* eversion can be produced quickly. There are no muscles within the *endophallus* and its inversion is caused by blood *haemolymph* pressure and air alone, in effect inflating it outside its body.

A single drone is capable of delivering up to 11 million *sperm* cells; the queen has a capacity to store about six million in her *spermatheca*, although she may well hold up to 60 million *sperm* after multiple mating. It is thought that the reason why the queen has so many mating partners is to help reduce the incidence of mating with a close relative.

It is also thought drones are able to recognise their own queen by smell, which will help deter them from mating with her, although this has yet to be proved conclusively.

There is strong evidence that the queen will return for further mating if somehow her *spermatheca* is insufficiently full.

Dissected reproductive organs

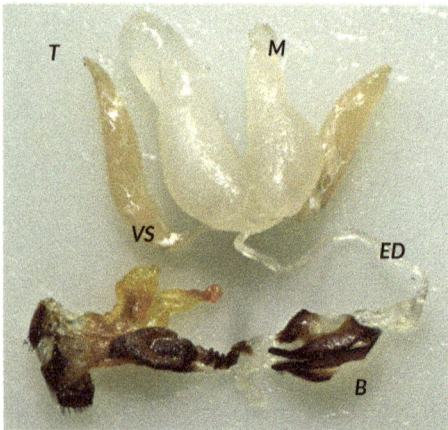

Figure 47: Dissected from the abdomen: Seminal ducts: SD. Testes: T residue is on the very top left, bulb. Vas deferens: VS. Mucus glands: MG. Ejaculatory duct: ED. See Drawing 8: Dade drawing plate 15.

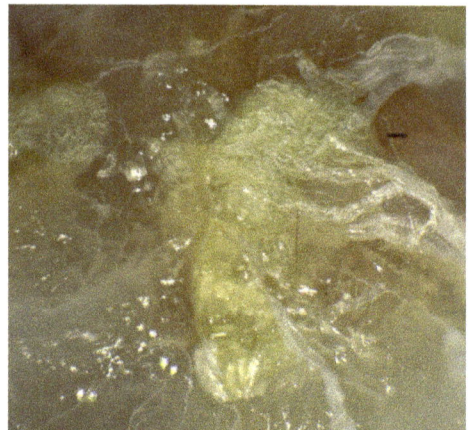

Figure 48: The residual testicle *of a mature drone showing the supporting* trachea *and* tracheoles.

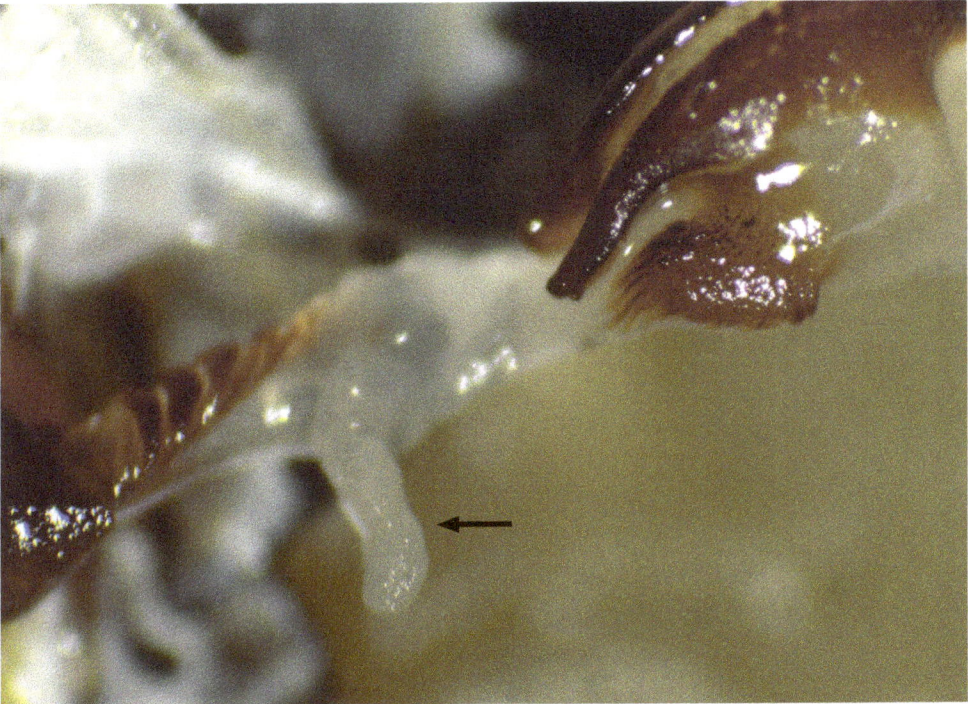

Figure 49: **Fimbriate** *lobe in uneverted state.*

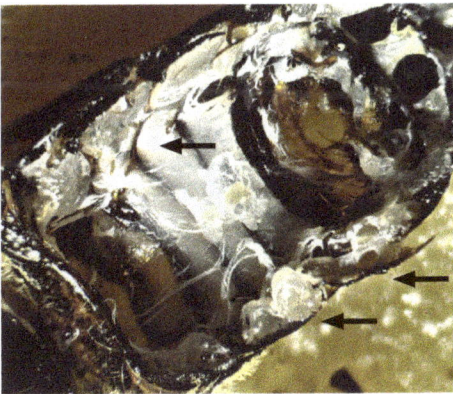

Figure 50: The extreme musculature on the side and bottom of the abdomen, these are used to cause the eversion of the endophallus.

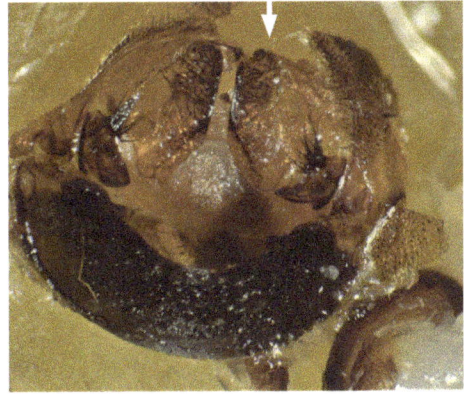

Figure 51: Dissected male opening. Showing the pair of rudimentary claspers. These are reduced in the honey bee genus, Apis: *They appear as two small sclerites attached to the sternites, they serve no function in the mating process unlike in other insects.*

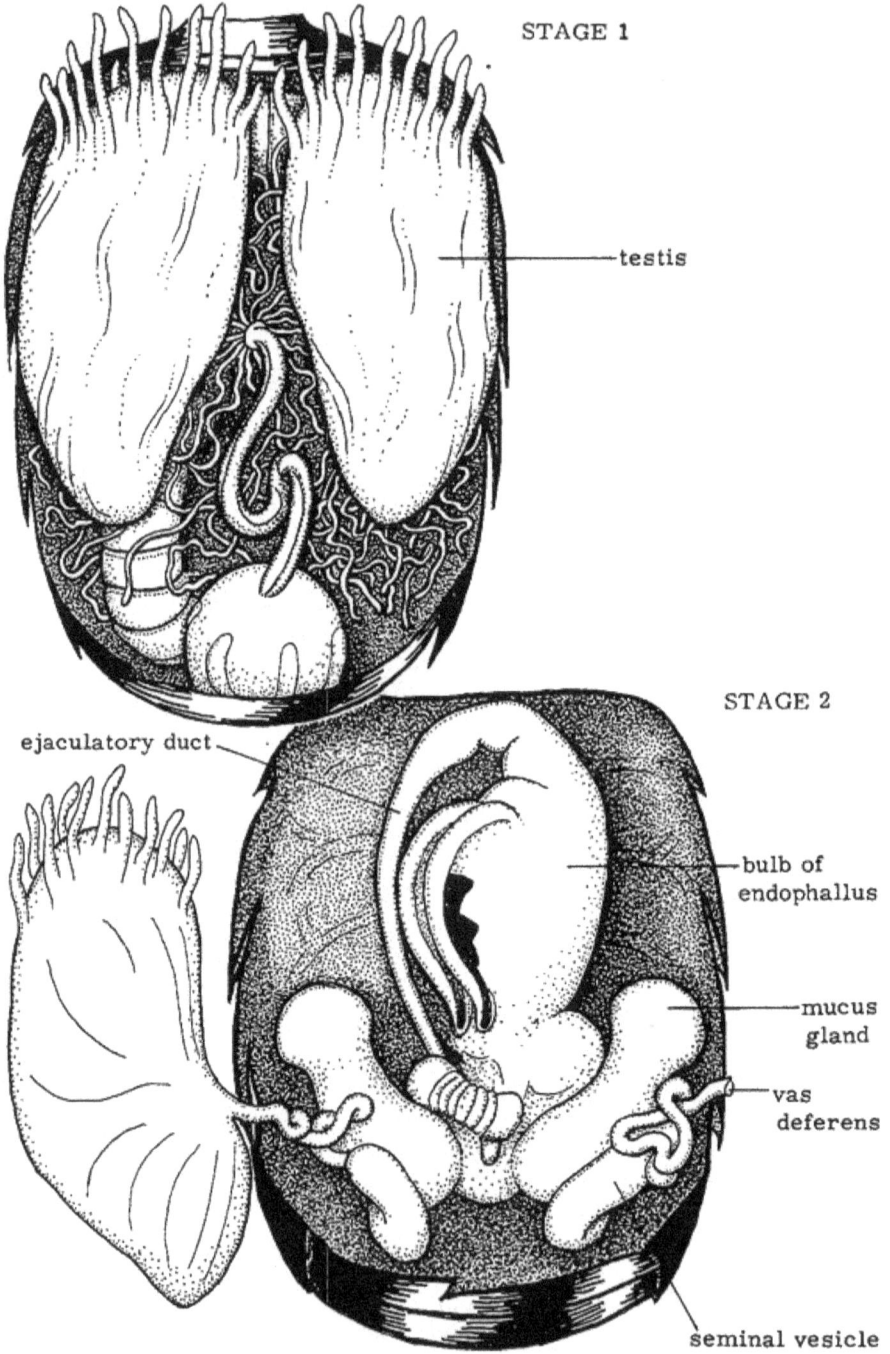

STAGE 1

testis

STAGE 2

ejaculatory duct

bulb of endophallus

mucus gland

vas deferens

seminal vesicle

Drawing 7: Plate 14: Dissection of immature newly emerged drone. Stage 1, viscera undisturbed. Stage 2, testes laid out and alimentary canal removed to expose the complete reproductive apparatus.

Taken from Dade's book: Anatomy and Dissection of the Honeybee. Kind permission to use them has been obtained from copyright holder IBRA.

testis

vas deferens

mucus gland

seminal vesicle

semen

mucus

bulb

A8, tg

dorsal plate
of bulb

lateral
plate

A9, st

horn

A8, st

bulb

clasper

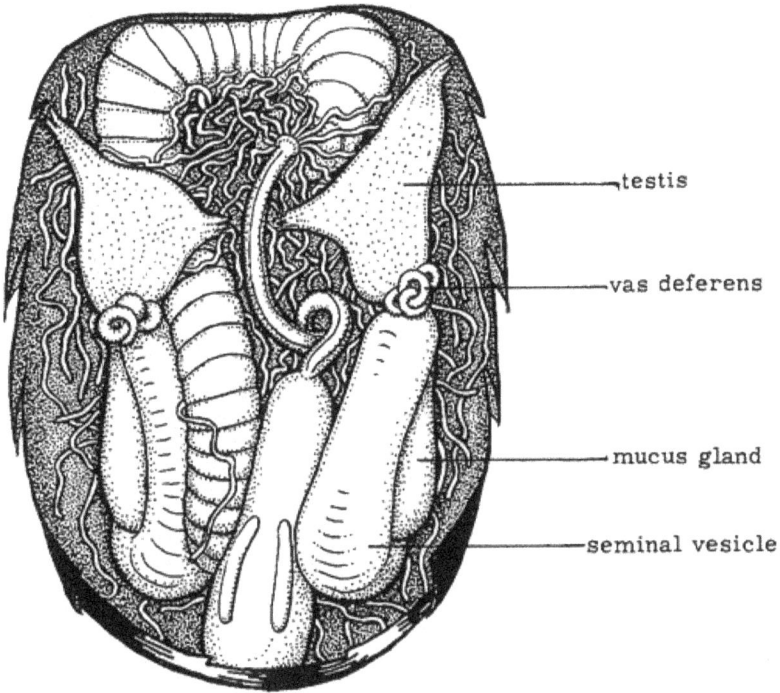

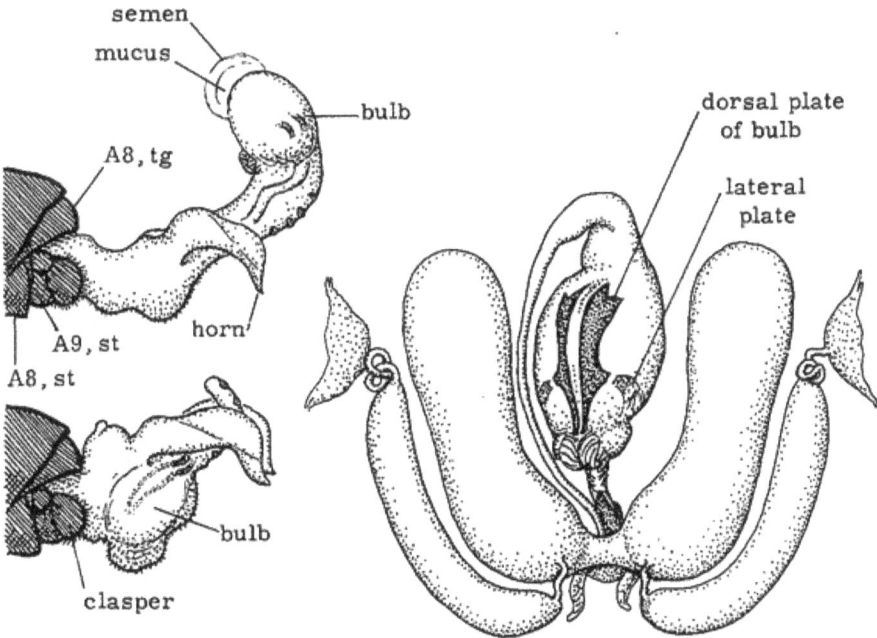

Drawing 8: Plate 15: Dissection of mature drone, viscera undisturbed. Below, right: the reproductive apparatus removed and laid out. Below, left: two stages in induced eversion of the endophallus.

Taken from Dade's book: Anatomy and Dissection of the Honeybee. Kind permission to use them has been obtained from copyright holder IBRA.

Honey bee genetics: The drone's role

A brief description of honey bee genetics setting out how the bees reproduce and why the drone has no father; the drones also have no sons, but at most, they have grandsons!

Chromosomes are long structures that are found in most cells. They contain the *genes* of an organism (humans have about 24,000 *genes*, bees have about 15,000 *genes*). Most animals normally have two sets of *chromosomes*: one set comes from the mother and one from the father.

They are called *diploid*. (Di means two, and ploid stands for *chromosome*) Human beings have 46 *chromosomes*: we get 23 from our mother's egg and 23 from our father's *sperm*. Bees have a different number of *chromosomes*. Females, workers and queens, have 32; 16 are contributed by the queen's eggs and 16 come from the drone's *sperm*.

Since drones hatch from unfertilized eggs, they only have the 16 *chromosomes* that were in the egg. Drones are *haploid* (from the Greek word haplos meaning single).

Haploid males are a characteristic of the order *Hymenoptera*, because they only have one set of *chromosomes*. The reproduction process in which the offspring develops from unfertilized eggs is called *parthenogenesis*, meaning asexual.

As the drone egg is unfertilised it can only carry half of the queen's 32 chromosomes, so she can only pass on half of her genes to her offspring. When the egg develops it divides into 2 parts, splitting the 32 chromosomes in half.

DNA

Adapted from National Human Genome Institute

cell

nucleus

chromosome

gene

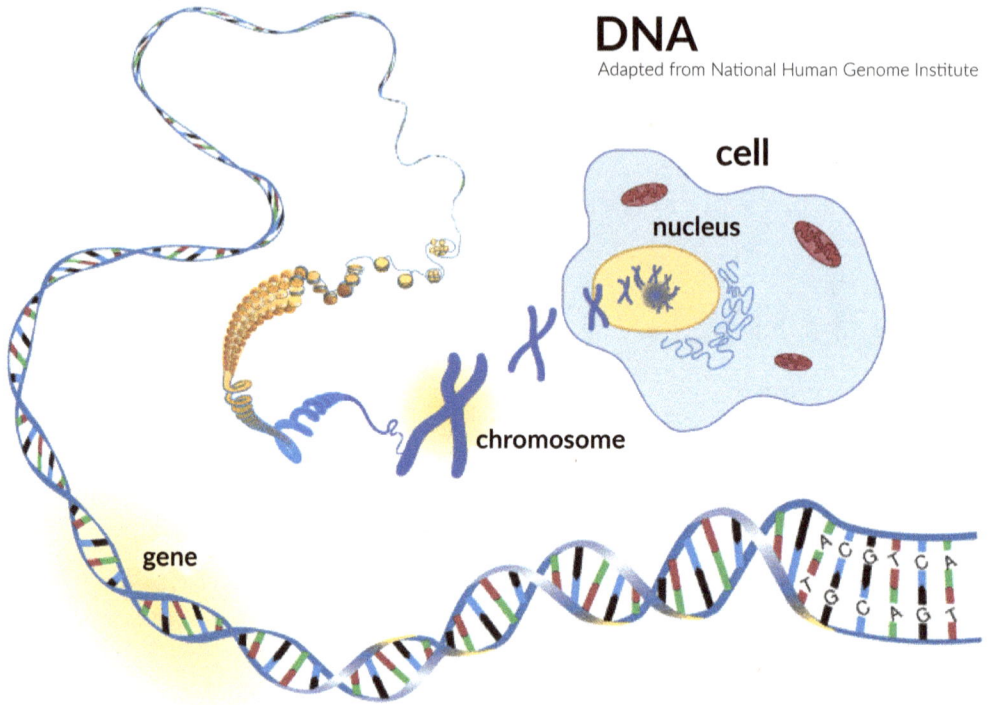

Figure 52: Reprinted by kind permission of Ryan Evans.

This process is called *meiosis*. Each egg contains a unique collection of her genes, so each egg is different. Drones, on the other hand, only have 16 *chromosomes* to begin with, so each *sperm* must contain all the genes of the drone. This means that each *sperm* from a single drone is identical. They are clones.

This is different from most other animals, where each *sperm* is unique, just as each egg is unique. If we use humans as an example, the females have two sex XX *chromosomes* both the same, and are called *Homozygous*, whereas males have two different sex *chromosomes* XY. This is called *Heterozygous*.

Depending on what sex *chromosome* the *sperm* cell has X or Y, determines the sex of human babies.

In bees, the female carries two copies of the same *genes* and are *homozygous*. However the drone has only one set of *genes* his genetic make-up is described as *hemizygous*.

Diploid drones can occur from a fertilised egg, according to Woyke (1963a). Professor Woyke in 1963 elaborated a method to rear *diploid* drone *larvae* to the *imago* stage.

A pair of *genes* for the same characteristic is called an *allelomorphic* pair, which is usually shortened to *alleles*.

Sex determination in the honey bee is controlled by about 15 sex *alleles*: coded for example: a, b, c, d, e, f, g, h. Normal drones developed from unfertilized eggs, with *allele* a, or b, or c, or d.........develop into *haploid* drones.

Fertilized eggs usually develop bee females (workers and queens), in which the sex determining *alleles* are different (ab, ac, ad, etc).

Drones that have two sets of *chromosomes* but the same version of the sex *allele* are in effect *haploid* having only one sex *allele* contributing genetic material where there should be two different. From these *homozygous* eggs, aa, bb, cc ... *diploid* drones can develop.

After the *diploid* drone *larvae* hatch from the eggs, they are not reared by the worker bees, but are eaten by the nurse bees within 6 hours after hatching. Professor Woyke in 1963 elaborated a method to rear *diploid* drone larvae to the *imago* stage. See figure 53.

The other odd thing about bees is their habit of multiple mating. The queen is quite promiscuous and mates with from 10 to 20 drones, usually in one or two mating flights over the course of a couple days. The *sperm* is stored for years in an organ called the *spermatheca*.

Actually, the *sperm* from one drone is more than enough to fill the *spermatheca*. Therefore, it seems the queen goes out of her way and takes great risks to mate with so many drones, just to create extra genetic diversity for her colony.

It is thought that one of the reasons why the queen mates with so many drones is due to her trying to obtain the rare alleles that are found amongst the drone groups, these could help to combat new diseases in the future.

Figure 53: Diploid drone, Woyke 1977.

In addition, since sub-families of worker bees (share a common father) tend to specialise in performing certain tasks in the hive, a diversity of fathers may enable the colony to perform more efficiently. There is well researched evidence that colonies headed by multiple-mated queens outperform single-mated queens with a dramatically better chance of winter survival.

The *phenotype* of an individual honeybee, a colony, or a population, is the set of *observable* characteristics such as size, colour, honey production, wintering ability and defensiveness. It is these characteristics that the bee breeder aims to alter. Longer term changes also occur but take time before a pattern becomes measurable.

The *genotype* (the "genetics") of a bee or colony is the set of inherited genetic instructions encoded in its DNA. These can be either dominant or recessive. A recessive allele does not become a trait unless both copies of the gene, one from the queen and drone, are present. If one dominant allele and one recessive allele are present, the dominant allele trait will be expressed. When we refer to two bees having different "genes," what we really mean is that they have two different forms (variants) of the same gene. Both in natural selection and in traditional selective breeding, selection is applied to the expression of the *phenotype*, rather than the *genotype*, since it is the *phenotype* that directly interacts with the environment, and is an observable characteristic.

Selection of the honeybee was not strongly influenced by humans because basic bee reproduction was not understood until about 1850 after Langstroth developed the movable frame hive. Suddenly beekeepers not only understood bee reproduction, they could also manipulate the hive and control the queen.

Within a colony, there are usually 7 to 10 *subfamilies* because of the queen mating with up to 20 different drones. Not all of the *sperm* from up to 20 matings is stored. Since all the *sperm* produced by a drone are genetically identical, each subfamily is composed of sisters that are more closely related than full sisters of other animals. Often called *super sisters*, they will have three-quarters of their *genes* in common.

Despite the complicated family structure, the basic principles of genetics still apply to bees. On rare occasions, a *gene* entering an egg or *sperm* has changed somewhat and will have a different effect than the original *gene*.

The process of change is called, *mutation*.

Mutations and their effects

More than 30 specific visible mutations have been described in bees. Generally, these mutations produce a striking effect. Known mutations affect the colour of, shape, and presence of eyes, the colour and hairiness of bodies, the shape and size of wings, and nest-cleaning behaviour. Most mutations are recessive, and are first observed in drones.

The reason for this is that they do not have paired *chromosomes*, only single, so no dominant *gene* can be preferentially expressed.

Problems in the hive

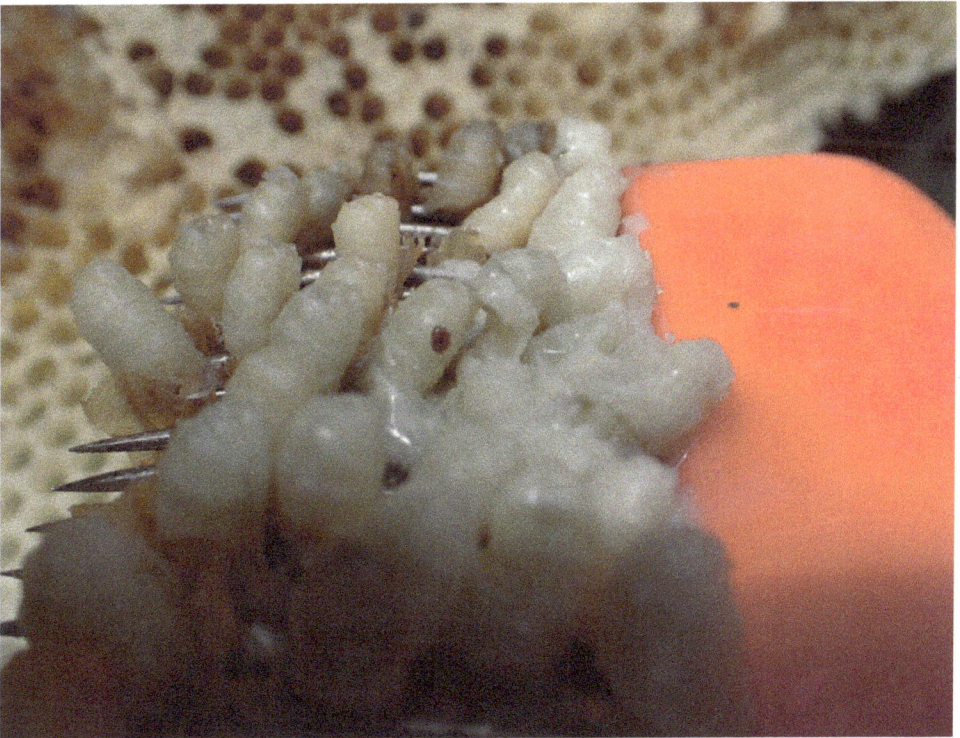

Figure 54: Drone removal as a means of treating Varroa mites.

A spiked fork is used to uncap the sealed drone cells: The logic of this method is based on the reproductive cycle of the *Varroa*, who prefer to lay eggs inside drone cells; the longer development time of 24 days allows for more *Varroa* developing.

While drone brood uncapping will remove a significant proportion of Varroa mites from the hive, it removes the genetic representation of that colony from the local gene pool until more drones can be raised to maturity, a period of over a month. It is therefore not approved of by beekeepers who are interested in genetic diversity and a strong drone population.

Removal, if done, often takes place at the peak of drone laying, setting back timing of drone flight when most queens are about. The more drones that are removed the more the queen produces to keep up the status quo, thus placing an additional burden on colony resources.

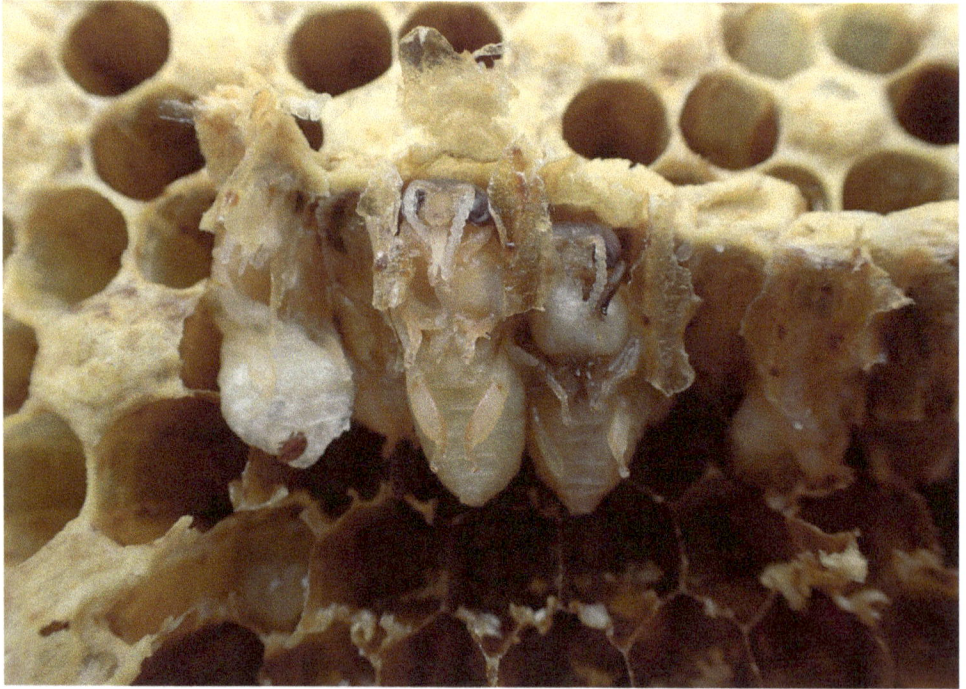

Figure 55: Drone capping removal.

The *Varroa* mite

A member of the spider family introduced to the UK over 20 years ago, it has become tolerant of some chemical treatments during this time.

Varroas pierce the bees' soft tissue between the abdominal *Tergite* plates and use the fat bodies stored underneath for food. Viruses are able to enter the bee's body through the wound which does not heal well, which may result in colony collapse.

Figure 56: The Varroa mite, anterior view.

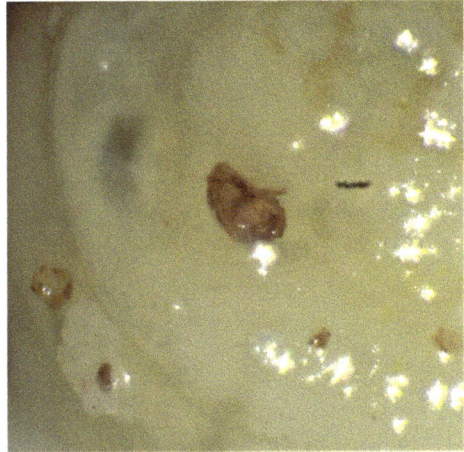

Figure 57: An immature Varroa on drone larvae body about 7 days old.

One line of current thinking is to leave the hives untreated and to allow natural selection to take place. This will take time and will result in many colony losses, but it will enable the bees to live in a tolerant relationship with the parasite.

As the bees have not encountered these pests before they are not adapted to handle them. The placing of hives next to each other can encourage *Varroa* infection owing to bee drift between the colonies.

In the wild, bee colonies would tend to be several kilometres apart so the risk would be minimal.

Figure 58: The Varroa feeding between plates. Compare the size difference between the bee and the Varroa.

Drone laying queen

If a virgin queen bee fails to mate or is infected with a virus she may become a drone laying queen. Sometimes if a queen is partially mated, or is running out of *sperm*, she will lay fertilised eggs for a period of time then unfertilised eggs thereafter.

The increasing proportion of drones will cause the colony to collapse as no workers are being produced.

The drones that develop at this stage are normally smaller and have stunted *abdomens* due to malnutrition, and also because they were reared in the smaller worker comb cells.

Scattered brood pattern

Another problem that occurs from bad mating is 'scattered brood patterning'. This is caused by the queen mating with a drone that is closely related to her resulting in a diploid drone being laid (see above).

The workers are able to recognise these by smell within a period of time and then remove them, leaving the resulting scattered brood pattern. If allowed to develop, or raised in an incubator away from worker policing, these drones would mainly be infertile. The solution to this problem is to remove the old queen and introduce a new well-mated one.

Drone laying worker

If the queen dies or swarms and the colony miss the opportunity to raise a new queen, one or several worker bees will often start to lay eggs as a last resort. These eggs are unfertilised as a worker does not mate, resulting in the hatching of drone larvae that could develop into drones. In this situation beekeepers often introduce some eggs or day old *larvae* in the comb from another colony and trust that the bees will be able to raise a new queen, or a new queen can be introduced,

Figure 59: Drones grubs and cells taken from a drone laying queen hive.

depending on how long they have been without one and how strong the colony is. Often it is best to amalgamate the queenless colony with a queenright one.

Drone eviction

Drones that do not mate live on average 21-32 days during spring to midsummer, but from late summer to autumn they can survive for up to 90 days. However, they will usually be removed from the hive by the workers before the winter, as food will be at a premium.

Evicted drones as a sign of colony health due to starvation

It is not uncommon in poor summers to find them evicted and crawling on the ground due to food shortages, particularly pollen. The colony must be queen right in order for them to be evicted.

If a virgin queen is present they will not be evicted until she starts to lay eggs. It is possible for a few drones to over winter if surplus food has been stored. Later they will be sacrificed for the benefit of the hive.

Disease transfer

There is a risk of drones transmitting disease to other colonies, as they seem to be tolerated as guests when visiting. A noted problem is the transmission of chronic bee paralysis virus to the queen, who in turns infects the colony. It is during mating that the *semen* is passed on to the queen and the virus hitches a lift in the *seminal fluid*. It has also been noted that *Nosema Apis* spores have been found in *semen* of older drones over 21 days. The possible reason for it not occurring sooner is due to the makeup and vitality of the *seminal fluid*, as older drones start to decline as they age.

Drone sperm viability due to exposure to pesticides in the colony

Recent studies have shown that during the development stages of the drone from egg to final emergence at 24 days, the presence of pesticides residue in the wax surrounding them can have a drastic effect on their *sperm* viability.

The pesticides vary from in-house treatments for *Varroa*, which are changed and continuously applied throughout the year, to those brought in by foraging bees, such as neonicotinoids. The outcome in comparison to drones reared on clean wax was pronounced.

The drones tested were all over 18 days old and had viable *sperm*, albeit compromised. Most drones are sexually mature at this age and not before. Samples were also taken from 10 day old drones and the evidence was the same, suggesting that the damage to the *sperm* happens much earlier in the capped cells where exposure is greatest.

The results showed a 20% drop in *sperm* viability from drones reared with pesticide residues in the colony. As a result any queen mated with a compromised drone will suffer reduced fertilised egg laying and a shortened fertile life. Prolonged use of old comb which stores pesticides over time can cause total sterilisation.

Drone infertility

Drones need a high protein diet, not only to grow but to help make *sperm*, a lack of protein at the *larval* stage, when the *testes* start their development, and again when they leave the cells at 24 days old, as the *sperm* continue to develop further, will be highly prejudicial. This can result in infertility and thus a poorly mated queen. A shortage of pollen is usually caused by poor weather conditions.

Drone visitations to other hives

Unlike worker bees drones seem to be able to visit and stay in other hives without being challenged by the guard bees. They even allow them to feed freely on stores; this might well be an evolutionary practice to allow the drones to refuel and ensure fertilisation of the virgin queens and not to rob the stores. On the downside drones could be a vector for disease transfer. It has been quoted that up to 15% of the drone population could be visitors. If a hive is found empty the drones will not stay.

Drone mutations

As described in the genetic section mutations are first seen in drones. One of the easiest to see is white-eyed drones. These are destined to spend their lives inside the hive as they are blind. This occurs, because the drones develop from *hemizygote* eggs. (A *diploid* individual with only one *allele* at a given locus or *gene* instead of the typical two).

Figure 60: A drone laying queen showing laying pattern and raised capping, often called pepperpotting.

Glossary

Abdomen. An area containing the digestive and reproductive organs.

Alleles. Alternative forms of a gene that arises by mutation and is found at the same place on a chromosome.

Antennae. Latin antemna, horizontal mast spar, the main sensory organ of the bee.

Bulb of endophallus; penis bulb

Bursa copulatrix. Mating pouch.

Bursal. Pouch

Cervix of endophallus. Neck of endophallus.

Chitinized plates. Hardened plates.

Chromosome. Found in every cell nucleus, divide into new sets.

Cornua. Horns found on the endophallus, coloured orange on mature drones.

Copula. Mating.

Cuticular intima. Cuticle coating.

Diploid. Di means two, and ploid stands for chromosome, having two sets.

Endophallus, meaning a 'penis held within' the word endo means within and phallus means penis

Ejaculatory duct. Tube leading from the seminal vessels to the bulb, conveys the sperm and mucus.

Eversion. The pushing out and turning inside out of the endophallus.

Flagellum. Latin for a little whip, the end section of the antennae.

Fissures in the outer layer of the cervix. A groove running along the neck.

Fimbriate lobe. A petal like structure, acting as a balloon when turned inside out.

Ganglion cells. A bunch of nerve cells that are found in seven different areas of the spinal cord, each acting as a separate control from the brain.

Gene. Unit of inheritance found in every chromosome.

Genotype. The genetics of a bee or colony a set of inherited genetic instructions encoded in its DNA

Gonopore. Opening of the ejaculatory duct;

Haploid. Greek word haplos meaning single, containing only one set of chromosomes.

Haemolymph. Blood of the bee.

Hemizygote. A diploid individual with only one allele at a given locus or gene instead of typical two.

Homozygous. Processing only identical set of chromosomes as the mother.

Imago .Adult honey bee, final stage of growth.

Instar. Phase between two stages of growth.

Lamina parameralis. Upper claspers, used in other insects when mating, too small to be of any use in bees.

Lateral oviduct. Top duct connecting to the ovaries.

Malpighian tubules. Excretory organs of the bee.

Mating sign. Part of the endophallus left behind in the queen's sting chamber.

Median oviduct. Duct where eggs pass through.

Mucous glands. Producers of mucus.

Mucus. Part of the final ejaculate.

Ommatidium. Individual structural lens of the compound eye.

Organ of Johnson. Found in the antennae used for determining air speed when in flight and sound detection.

Ocelli. Simple eye.

Parthenogenesis. Sexual reproduction without fertilisation.

Pedicel. A small stalk like structure, part of the antennae.

Phallotreme. Opening for the endophallus and anus.

Phenotype of an individual honeybee, a colony, or a population, is the set of observable characteristics

Pheromone. Chemical messages that are produced by the bee to communicate.

Polarised light. A light that is vibrating in more than one plane, not visible to humans.

Post-eclosion. After hatching from the cell.

Pneumophyses of endophallus. Lateral lobes.

Primordial germ cells. Biological cell that gives rise to the gametes of an organism that reproduces sexually.

Proctodaem. Hind gut.

Pupae. Stage in the development when complete metamorphosis between larvae and egg.

Sac of the bulb. Orifice.

Scape. Rigid stalk, connecting with the head joint.

Semen. Mixed ejaculatory fluid

Seminal vesicle; vesicula seminalis. Sperm storage vessel.

Sensilla picola. Specialised individual sense organs

Sclerites. Outer covering of the bee, its exoskeleton made up of chitin.

Sternite. Plates covering the abdomen.

Spermatheca. Queen's sperm storage organ.

Spermatogenesis. The process of sperm development.

Stomodaeum. Fore gut.

Symbiotic relationship. Mutual benefit of both parties.

Testes. Latin for witness. Organ where sperm are made.

Thorax. Torso.

Vasa deferentia. Vessels that convey sperm.

Ventriculus. The mid gut and true stomach where digestion takes place.

Vestibulum at base of the endophallus .A large internal cavity.

References

Understanding Bee Anatomy: A full colour guide by Ian Stell

The Buzz about Bees Biology of a Superorganism: Jürgen Tautz

Pheromones of Social Bees: John B Free

Queen Bee: Biology, Rearing and Breeding: David Woodward

Beekeeping study notes: Yates and Yates.

The Anatomy of the Honey bee: Snodgrass.

Honey bee anatomy and dissection: Dade.

Age-specific olfactory attraction between Western honey bee drones (*Apis mellifera*) and its chemical basis. PLoS ONE 12(10): e0185949. https://doi.org/10.1371/journal.pone. 0185949

Woyke J. (1963) Drones from fertilized eggs and biology of sex determination in the honeybee. Bulletin de L'de Academie Polonaise des Sciences Cl. V. - Vol. XI, No. 5, Serie des sciences biologiques
https://www.researchgate.net/publication/257214500

Woyke J. (1963a): What happens to diploid drone larvae in a honeybee colony. Journ. apic. Res. 2(2): 73-76
https://www.researchgate.net/publication/239521613

Woyke J. (1963b) Rearing and viability of diploid drone *larvae*. Jour. apic. Res. 2(2): 77-84)
https://www.researchgate.net/publication/239521612

Woyke Jerzy (1984) Ultra structure of single and multiple *diploid* honeybee *spermato*-zoa. Journ. apic. Res. 23(3):123-135.
https://www.researchgate.net/publication/239521691

Woyke Jerzy (2008): Why the eversion of the *endophallus* of honey bee drone stops at the partly everted stage and significance of this. Apidologie, Springer Verlag, 2008, 39 (6), pp.627-636. Hal-00891976
https://www.researchgate.net/publication/249998814

Woyke Jerzy (2011) Mating sign of queen bee originates from two drones and the process of multiple mating in honey bees. Journ. apic. Res. 50 (4): 272-283 PDF 6.4 MB
https://www.researchgate.net/publication/249998814

Woyke Jerzy (2016) Not the honey bee (*Apis mellifera* L.) queen, but the drone determines the termination of the nuptial flight and the onset of oviposition - the polemics, abnegations, corrections and supplements Journal of Apicultural Sciences 60 (2): 25-40
https://www.researchgate.net/publication/31162420

Woyke J. Ruttner F. (1958): An anatomical study of the mating process in the honeybee. Bee World 39(1): 3-18 https://www.researchgate.net/publication/257415271

Faten Abdelkader, Guillaume Kairo, Sylvie Tchamitchian, Marianne Cousin, Jaques Senechal, et al. *Semen* quality of honey bee drones maintained from emergence to sexual maturity under laboratory, semi-field and field conditions. Apidologie, Springer Verlag, 2014, 45(2), pp.215-223. 10.1007/s13592013-0240-7

HAL-01234717 Exposure to pesticides during development negatively affects honey bee (*Apis Mellifera*) drone *sperm* viability Adrian Fisher II Juliana Rangel https://doi.org/10.1731/journal.pone.0208630

Sexual selection in *Apis* bees Apidologie 36 (2005)   187-200

Comparative anatomy of male genital organs in the genus *Apis* G & N Koeniger Apidologie Springer Verlag 1991 .22 (5) pp 539-552 hal-00890856

Sex and Caste-Specific Variation in Compound Eye Morphology of Five Honeybee Species PLoS One. 2013; 8(2): e57702

Coelho: The effect of thorax temperature on force production during tethered flight in honeybee (*Apis mellifera*) drones, workers, and queens. Physiological Zoology 6 4(3).823 -835.

Acknowledgements

A huge debt of gratitude must go to Professor Jerzy Woyke for kindly allowing me to use his research data and drawings and for reviewing the contents of this book.

IBRA has the copyright for Dade's Anatomy & Dissection of the Honeybee. They also hold the copyright to Bee World articles. Permission to use both sets of data has been kindly granted and acknowledged.

My external thanks go to all the bee scientists who have freely published their research. The many beekeepers who have educated me in a multitude of ways, especially Chris Utting, Julia Elkin, Paul Smith, David Raitt, Lilah Killock and Alan White for all those bee thoughts over a beer or two. My wife Catherine Kingham the queen in my life.

I am in constant debt to Pat and David Woodward who are both highly qualified microscopists, who have helped and advised me over the years and have always encouraged me to pursue my interests, a very big thank you.

I reserve special thanks for Richard Simpson for his help in editing, suggestions and advice, as well as the time spent within the drone congregation areas, enriching my knowledge even further.

Front cover watercolour illustration by Eunike Nugroho who has kindly given permission for use for education purposes. Eunike is a botanical and natural history artist based in Indonesia; she has exhibited widely and has a wide portfolio of work in print. To see more of her fantastic work, please visit www.eunikenugroho.com. The watercolour painting was inspired by a photograph of a drone taken by Guillaume Pelletier in Canada who has kindly given permission to use.

For those of a technical mind the microscopes used were:

- Meiji compound biological microscope MX4200H

- Brunel zoom dissecting trinocular stereo microscope BMDZ. Brunel has a vast range of quality microscopes new and used suitable for every pocket. They offer outstanding service and advice on all matters concerning the microscope world. *www.brunelmicroscopes.co.uk*

Cameras used:

- Toupcam E3CMOS. 12 mega pixels. USB 3. Software: ToupView.

- Olympus OM D E-M1 camera with 60mm macro lens.

All photographs by Graham Kingham unless otherwise noted.

Index

www.ingramcontent.com/pod-product-compliance
Lightning Source LLC
Chambersburg PA
CBHW042108210326
41519CB00064B/7587